Ordinary and Partial Differential Equations for the Beginner

Ordinary and Partial Differential Equations for the Beginner

László Székelyhidi
University of Debrecen, Hungary

NEW JERSEY • LONDON • SINGAPORE • BEIJING • SHANGHAI • HONG KONG • TAIPEI • CHENNAI • TOKYO

Published by

World Scientific Publishing Co. Pte. Ltd.
5 Toh Tuck Link, Singapore 596224
USA office: 27 Warren Street, Suite 401-402, Hackensack, NJ 07601
UK office: 57 Shelton Street, Covent Garden, London WC2H 9HE

Library of Congress Cataloging-in-Publication Data
Names: Székelyhidi, László.
Title: Ordinary and partial differential equations for the beginner / by László Székelyhidi (University of Debrecen, Hungary).
Description: New Jersey : World Scientific, 2016. | Includes bibliographical references and index.
Identifiers: LCCN 2016016298| ISBN 9789814723985 (hardcover : alk. paper) | ISBN 9789814723992 (pbk. : alk. paper)
Subjects: LCSH: Differential equations--Textbooks. | Differential equations, Partial--Textbooks.
Classification: LCC QA371 .S954 2016 | DDC 515/.352--dc23
LC record available at https://lccn.loc.gov/2016016298

British Library Cataloguing-in-Publication Data
A catalogue record for this book is available from the British Library.

Printed in Singapore

To Kati

Contents

Preface

Differential equations play a central role in mathematics, both in theory and applications. In this volume, we present the material which has been used by the author when teaching at MSc university level. Besides the standard subjects on ordinary differential equations, we also include those parts of the theory of first and second order partial differential equations, where the theory and solution methods are closely related to ordinary differential equations.

The first five sections of this book are devoted to the theory of existence and uniqueness of the solutions of ordinary differential equations and initial value problems. We also consider parametric problems. However, separate sections exhibit practical solution methods for different types of equations. In particular, we present some useful methods to solve implicit differential equations, finding exponential polynomial solutions, solving boundary value problems for second order linear equations, and applying power series and Laplace transform to find particular solutions of differential equations and initial value problems.

Two sections deal with first order partial differential equations and Cauchy problems. In particular, we present the basics of the theory of characteristics and its use for solving linear and nonlinear first order partial differential equations and related Cauchy problems.

The last three sections include classical methods to solve some problems concerning higher order partial differential equations. Here the main emphasis is not on the theory but on the different solution methods.

Each section is completed with a number of problems and exercises.

Answers to selected problems can be found in the last chapter.

In this work, we extensively used the following books and papers: [Kamke (1944, 1946); Coddington and Levinson (1955); Pontryagin (1962); Stepanow (1963); Hartman (1964); Petrovski (1966); Vladimirov (1974); Székelyhidi (1985, 1986); Vladimirov (1988); Boyce and DiPrima (2001); Filippov (2000); Arnold (2006); Székelyhidi (2006); Filippov (2007); Logan (2011); Schroers (2011)].

L. Székelyhidi

Chapter 1

ORDINARY DIFFERENTIAL EQUATIONS

1.1 Basic concepts and terminology

In this work, by an *interval* we mean a real interval containing at least two points. The set of the natural, real, respectively complex numbers will be denoted by $\mathbb{N}$, $\mathbb{R}$, respectively $\mathbb{C}$, and $\mathbb{R}_+$ denotes the set of positive real numbers.

The equation

$$x' = f(t, x) \tag{1.1}$$

will be called an *ordinary vector-differential equation of the first order (shortly: first order differential equation)* on the set Ω, where Ω is a non-empty open subset in $\mathbb{R} \times \mathbb{R}^n$ and $f : \Omega \to \mathbb{R}^n$ is a given function. The equation

$$x(t_0) = x_0 \tag{1.2}$$

will be called an *initial condition* for (1.1), where (t_0, x_0) is an element of Ω. The system (1.1),(1.2) will be called a *Cauchy problem*, or an *initial value problem* for (1.1), and (t_0, x_0) is called the *initial data*. The set Ω is called the *domain*, and the function f is called the *right side* of the differential equation (1.1).

It is often useful to write (1.1) and (1.2) in coordinate form. We then have

$$\begin{aligned}
x_1' &= f_1(t, x_2, x_2, \ldots, x_n) \\
x_2' &= f_2(t, x_2, x_2, \ldots, x_n) \\
&\cdots \\
x_n' &= f_n(t, x_2, x_2, \ldots, x_n)\,,
\end{aligned}$$

where we use the notation $x = (x_1, x_2, \dots, x_n)$ and $f = (f_1, f_2, \dots, f_n)$.

The function φ will be called a *solution of* (1.1) *on* I, if I is a non-void open interval in $\mathbb{R}$, further $\varphi : I \to \mathbb{R}^n$ is a differentiable function for which the point $\big(t, \varphi(t)\big)$ belongs to Ω, and

$$\varphi'(t) = f\big(t, \varphi(t)\big)$$

holds for each t in I. We emphasize that solutions of differential equations will almost always be meant on open intervals. We say that the solution φ of (1.1) *satisfies the initial condition* (1.2), if the point t_0 belongs to I and $\varphi(t_0) = x_0$. In this case, we say that φ *is a solution of the Cauchy problem* $(1.1), (1.2)$ *on the interval* I.

We call φ a *complete solution* of equation (1.1), if φ is a solution of (1.1), and (1.1) has no solution which is a proper extension of φ. We define the complete solution of the Cauchy problem (1.1), (1.2) in a similar, obvious manner.

Sometimes we shall use in this work the slightly loose concept of *general solution* of a given differential equation. By this we mean a family of functions, usually described by an algebraic formula which has the property that every member of this family is a complete solution of the differential equation, and every complete solution of the differential equation is included in the family. We shall see that in a wide class of differential equations a general solution can be given explicitly.

If in (1.1) we have $\Omega = I \times \mathbb{R}^n$, where I is a non-void open interval in $\mathbb{R}$, moreover $f : \Omega \to \mathbb{R}^n$ is a continuous function which is linear in its second variable, then we call (1.1) a *linear ordinary differential equation of the first order*, or shortly a *linear differential equation* on I. Hence the general form of linear differential equations is

$$x' = A(t)x + b(t)\,, \tag{1.3}$$

where $A : I \to L(\mathbb{R}^n)$ and $b : I \to \mathbb{R}^n$ are continuous functions. Here $L(\mathbb{R}^n)$ denotes the space of linear operators of the linear space $\mathbb{R}^n$. Using the standard basis of $\mathbb{R}^n$ the space $L(\mathbb{R}^n)$ can be identified with the space of all real $n \times n$ matrices. In this case, (1.3) can be written in the form

$$\begin{aligned}
x_1' &= a_{1,1}(t)x_1 + a_{1,2}(t)x_2 + \cdots + a_{1,n}(t)x_n + b_1(t) \\
x_2' &= a_{2,1}(t)x_1 + a_{2,2}(t)x_2 + \cdots + a_{2,n}(t)x_n + b_2(t) \\
&\dots \\
x_n' &= a_{n,1}(t)x_1 + a_{n,2}(t)x_2 + \cdots + a_{n,n}(t)x_n + b_n(t)\,,
\end{aligned}$$

where $A = \left(a_{i,j}\right)_{i,j=1}^{n}$ and $b = (b_1, b_2, \ldots, b_n)$.

The linear differential equation (1.3) is called *inhomogeneous*, if b is non-zero, otherwise it is called *homogeneous*. In general, given (1.3) the homogeneous linear differential equation

$$x' = A(t)x \tag{1.4}$$

is called the *homogeneous equation corresponding to* (1.3).

The differential equation (1.1) is called *autonomous*, if the right hand side is independent of t. In other words, the general form of an autonomous differential equation is

$$x' = f(x)\,, \tag{1.5}$$

where $f : D \to \mathbb{R}^n$ is a function, and D is an open subset in $\mathbb{R}^n$.

The basic problems concerning differential equations are *existence*, *uniqueness* and *representation* of the solutions. The problem of existence means that we are looking for reasonable conditions which are necessary, and/or sufficient for the existence of at least one solution of (1.1), or (1.1), (1.2). These conditions should refer to f and (t_0, x_0), that is the given data, obviously. The uniqueness problem means to find necessary and/or sufficient conditions on the given data for the existence of at most one solution. We remark that if a solution of (1.1) is given, defined on some open interval, then its restriction to any proper open subinterval is a solution, too, and it is different. The same holds concerning the Cauchy problem (1.1), (1.2), if we restrict a solution to any proper open subinterval containing t_0. This shows that to obtain uniqueness results we have to modify our expectation on uniqueness in a reasonable way. One possible way is to speak about uniqueness of a complete solution, only. Another way is to introduce the following concept. We say that the solution of the Cauchy problem (1.1), (1.2) is *locally unique*, if any two solutions coincide on the intersection of their domains. Using any of these ideas uniqueness cannot be violated by simply restricting a solution to a smaller interval. Finally, representation of the solutions means that we are looking for methods, theoretical or numerical, which can be used to compute the explicit form of the exact solutions. We shall see that there are quite general existence and uniqueness results for wide classes of differential equations and Cauchy problems, but a representation of all solutions is available only for rather special classes of equations. In the applications, one prefers numerical methods to approximate the solutions with a prescribed error. In this work, we are dealing with exact methods only, however, in some cases we point out how some theoretical results can be utilized to construct effective numerical methods. Also, in addition, in some sections we shall give practical computational methods to find solutions without going into theoretical details. These sections should be considered as "recipes" for solving problems about differential equations.

1.2 Problems

1. Give an example for a differential equation of the form (1.1) which has no solution.
2. Give an example for a differential equation of the form (1.1) which has more than one solution.
3. Give an example for a differential equation of the form (1.1) which has infinitely many solutions.
4. Give an example for a Cauchy problem of the form (1.1), (1.2) which has no solution.
5. Let $n = 1$, $\Omega = \mathbb{R} \times \mathbb{R}$ and $f(t, x) = \sqrt{|x|}$ for each (t, x) in Ω. Show that the corresponding Cauchy problem
$$x' = |x|$$
$$x(t_0) = x_0$$
has a unique complete solution for each t_0, x_0 in $\mathbb{R}$, if $x_0 \neq 0$, and it has infinitely many solutions on any open interval containing t_0, if $x_0 = 0$.
6. Let $n = 1$, $\Omega = \mathbb{R} \times \mathbb{R}$ and $f(t, x) = x$ for each (t, x) in Ω. Show that the function $t \mapsto x_0 e^{t-t_0}$ is the unique complete solution of the Cauchy problem
$$x' = x \tag{1.6}$$
$$x(t_0) = x_0 \tag{1.7}$$
for each t_0, x_0 in $\mathbb{R}$. (*Hint*: show that for each solution $\varphi : I \to \mathbb{R}$ of the differential equation $x' = x$ the function $t \mapsto \varphi(t)e^{-t}$ is constant.)
7. Let I be a real open interval, and $p : I \to \mathbb{R}$ is a continuous function. Show that if $\varphi_1, \varphi_2 : I \to \mathbb{R}$ are solutions of the differential equation $x' + p(t)x = 0$ on I, and φ_2 is non-vanishing, then $\frac{\varphi_1}{\varphi_2}$ is constant.
8. Suppose that $\varphi : I \to \mathbb{R}^n$ is a solution of the autonomous differential equation (1.5), and t_0 is in $\mathbb{R}$. Show that the function $t \mapsto \varphi(t + t_0)$ is a solution of (1.5) on the interval $I - t_0 = \{t - t_0 : t \text{ is in } I\}$.
9. Find all complete solutions of the following autonomous differential equations:

i) $x' = \sqrt{x}$
ii) $x' = e^{-2x}$
iii) $x' = 1 + x^2$
iv) $x' = \frac{x}{1+x^2}$
v) $x' = e^{x^2}$.

1.3 Auxiliary results from functional analysis

We recall that if (X, d) is a metric space, then we call the mapping $A : X \to X$ a *contraction*, if there exists a number q in the open interval $]0, 1[$ such that

$$d\big(A(x), A(y)\big) \leqslant qd(x, y)$$

holds for all x, y in X. We say that the point x in X is a *fixed point* of the mapping $A : X \to X$, if $A(x) = x$ holds.

Theorem 1.3.1. *(Banach's Fixed Point Theorem) Any contraction on a complete metric space has a unique fixed point.*

Proof. Let (X, d) be a complete metric space and $A : X \to X$ a contraction, that is, there is a number q in the interval $]0, 1[$ such that

$$d\big(A(x), A(y)\big) \leqslant q\, d(x, y)$$

holds for all x, y in X. Let x_0 be an arbitrary element of X and we define

$$x_{n+1} = A(x_n)$$

for each natural number n. We show that $(x_n)_{n\in\mathbb{N}}$ is a Cauchy sequence. Indeed, if $m \geqslant n$ are natural numbers, then

$$d(x_n, x_m) = d\big(A^n(x_0), A^m(x_0)\big) \leqslant q^n d(x_0, x_{m-n})$$

$$\leqslant q^n \big(d(x_0, x_1) + d(x_1, x_2) + \cdots + d(x_{m-n-1}, x_{m-n})\big)$$

$$\leqslant q^n d(x_0, x_1)\big(1 + q + q^2 + \cdots + q^{m-n-1}\big)$$

$$\leqslant \frac{q^n}{1-q} d(x_0, x_1),$$

which implies the statement, as $q^n \to 0$.

By completeness, the sequence $(x_n)_{n\in\mathbb{N}}$ converges to an element x in X. As obviously any contraction is continuous, hence the sequence $\big(A(x_n)\big)_{n\in\mathbb{N}}$

converges to $A(x)$. On the other hand, by $A(x_n) = x_{n+1}$, this latter sequence is a subsequence of $(x_n)_{n\in\mathbb{N}}$, hence its limit is x, which immediately implies $A(x) = x$, therefore x is a fixed point of the contraction A.

The uniqueness follows from the simple observation that if y is another fixed point of A, then we have

$$d(x, y) = d\big(A(x), A(y)\big) \leqslant q\, d(x, y)\,,$$

which is impossible, by $q < 1$, unless $d(x, y) = 0$, that is $x = y$. □

Let $[a, b]$ be an arbitrary closed real interval and let the sequence of functions $f_m : [a, b] \to \mathbb{R}^n$ $(m = 0, 1, \ldots)$ be given. We say that this sequence of functions is *uniformly bounded*, if there is a real number K such that

$$\|f_m(t)\| \leqslant K$$

holds for each t in $[a, b]$ and for all $m = 0, 1, \ldots$. We say that this sequence of functions is *equicontinuous*, if for every $\varepsilon > 0$ there exists a $\delta > 0$ such that for any t, s in $[a, b]$ with $|t - s| < \delta$ we have

$$\|f_m(t) - f_m(s)\| < \varepsilon$$

for all $m = 0, 1, \ldots$.

Theorem 1.3.2. *(Arzelà–Ascoli) Every uniformly bounded and equicontinuous sequence of functions with values in $\mathbb{R}^n$ on a closed real interval has a uniformly convergence subsequence.*

Proof. Let the sequence of functions $f_m : [a, b] \to \mathbb{R}^n$ $(m = 0, 1, \ldots)$ be uniformly bounded and equicontinuous. Let $(t_k)_{k\in\mathbb{N}}$ denote the sequence of rational numbers in the interval $[a, b]$. We consider the sequence $\big(f_m(t_1)\big)_{m\in\mathbb{N}}$. It is bounded, hence it has a convergent subsequence: $\big(f_{m_i}(t_1)\big)_{i\in\mathbb{N}}$. We renumber the corresponding functions: $f_i^{(1)} = f_{m_i}$ $(i = 0, 1, \ldots)$. Here $m_i \geqslant i$. This means that the values of the sequence of functions at the point t_1 form a convergent sequence. Now, we consider the sequence of the values of this sequence of functions at the point t_2. This is bounded again, hence it has a convergent subsequence. Repeating the previous argument, we construct a new sequence of functions $\big(f_m^{(2)}\big)_{m\in\mathbb{N}}$ which is a subsequence of the original sequence of functions and the sequences of its function values at t_1 and t_2 are convergent. Continuing this process after k steps, we obtain the sequence of functions $\left(f_m^{(k)}\right)_{m\in\mathbb{N}}$ which is a subsequence of the original sequence of functions and it has the property that the limits

$$\lim_{m\to\infty} f_m^{(k)}(t_i)$$

exist for $i = 1, 2, \ldots, k-1$.

Now, we consider the sequence of functions $\left(f_m^{(m)}\right)_{m\in\mathbb{N}}$. By the construction, this is a subsequence of the original sequence of functions and it is convergent at the points t_k with k in $\mathbb{N}$. We show that this sequence is uniformly convergent on the whole interval $[a, b]$. Let $g_m = f_m^{(m)}$ and let $\varepsilon > 0$ be arbitrary. By equicontinuity, there exists a $\delta > 0$ such that if t, s are in $[a, b]$ with $|t - s| < \delta$, then

$$\|g_m(t) - g_m(s)\| < \frac{\varepsilon}{3}$$

holds for $m = 0, 1, \ldots$. We choose a natural number p such that the distances of the neighboring pairs from the numbers $t_0 = a, t_1, t_2, \ldots, t_p, t_{p+1} = b$ are less than δ. Finally, we choose a natural number N such that

$$\|g_{m+k}(t_i) - g_m(t_i)\| < \frac{\varepsilon}{3}$$

holds for $m \geqslant N$, $k \geqslant 0$, $i = 0, 1, 2, \ldots, p+1$.

Let t be an arbitrary point in $[a, b]$. By the property of p, there exists a point t_r with $0 \leqslant r \leqslant p+1$ and $|t - t_r| < \delta$. Therefore

$$\|g_m(t) - g_{m+k}(t)\| \leqslant \|g_m(t) - g_m(t_r)\| + \|g_m(t_r) - g_{m+k}(t_r)\|$$

$$+ \|g_{m+k}(t_r) - g_{m+k}(t)\| < \frac{\varepsilon}{3} + \frac{\varepsilon}{3} + \frac{\varepsilon}{3} = \varepsilon\,,$$

and, by Cauchy's Convergence Principle, the uniform convergence of the sequence of functions $(g_m)_{m\in\mathbb{N}}$ – which is a subsequence of the original sequence – follows on the interval $[a, b]$. □

1.4 Problems

1. Give an example for a metric space X and a contraction $A : X \to X$ which has no fixed point.
2. Give an example for a metric space x and a mapping $A : X \to X$ satisfying $d\big(A(x), A(y)\big) < d(x, y)$ for each x, y in X which has no fixed point.

3. Let X be a compact metric space and $A : X \to X$ a mapping satisfying $d\big(A(x), A(y)\big) < d(x, y)$ for each x, y in X. Show that A has a unique fixed point.
4. Give an example of a uniformly bounded sequence of continuous functions on a closed real interval with values in $\mathbb{R}^n$ which has no uniformly convergent subsequence.
5. Give an example of an equicontinuous sequence of functions on a closed real interval with values in $\mathbb{R}^n$ which has no uniformly convergent subsequence.

1.5 Approximate solutions. Peano's theorem

We say that φ is an *ε-solution* of the Cauchy problem (1.1), (1.2) on J, if J is an interval in $\mathbb{R}$, $\varepsilon > 0$ is a real number, $\varphi : J \to \mathbb{R}^n$ is a continuous function having the property that $\big(t, \varphi(t)\big)$ belongs to Ω for each t in J, t_0 belongs to J, $\varphi(t_0) = x_0$, further φ is differentiable at all but finitely many points of J, and if φ is differentiable at some point t in J, then we have

$$\|\varphi'(t) - f(t, \varphi(t))\| < \varepsilon .$$

We point out that here J can be any type of real interval: finite or infinite, open or closed, or half open. If J contains either of its endpoints, then we suppose one sided differentiability at that point.

Lemma 1.5.1. *Let $\varphi : [a, b] :\longrightarrow \mathbb{R}^n$, $\psi : [a, b] \to \mathbb{R}$ be continuous functions having right sided derivatives at each point of the interval $[a, b[$. If for every t in $[a, b[$, we have*

$$\|\varphi'_+(t)\| \leqslant \psi'_+(t) ,$$

then

$$\|\varphi(b) - \varphi(a)\| \leqslant \psi(b) - \psi(a)$$

holds.

Proof. Let $\varepsilon > 0$ be given and we denote by I the set of all points y in $[a, b]$ for which $a \leqslant x \leqslant y$ implies

$$\|\varphi(x) - \varphi(a)\| \leqslant \psi(x) - \psi(a) + \varepsilon(x - a) .$$

Obviously, I is a closed interval and a is in I. Let $c = \sup I$. Suppose that $c < b$. As

$$\|\varphi'_+(c)\| \leqslant \psi'_+(c) ,$$

this implies $\varphi'_+(c) = \psi'_+(c)u$ with some u in $\mathbb{R}^n$ satisfying $\|u\| \leqslant 1$. On the other hand, the right sided derivative of the function $\varphi - \psi \cdot u$ at c is 0, hence there is a y with $c < y \leqslant b$, and if $c \leqslant x \leqslant y$, then

$$\|\varphi(x) - \varphi(c) - \big(\psi(x) - \psi(c)\big)u\| \leqslant \varepsilon(x - c)$$

holds which implies

$$\|\varphi(x) - \varphi(c)\| \leqslant \psi(x) - \psi(c) + \varepsilon(x - c)\,,$$

contradicting the definition of c. Hence $c = b$ and the lemma is proved. □

Lemma 1.5.2. *Let I be an interval in $\mathbb{R}$, H an open set in $\mathbb{R}^n$, further let $f : I \times H \to \mathbb{R}^n$ and $u_m : I \to H$ $(m = 0, 1, \dots)$ be continuous functions. If the sequence of functions $(u_m)_{m\in\mathbb{N}}$ is uniformly convergent on every compact subset of I, then the sequence of functions $t \mapsto f\big(t, u_m(t)\big)$ $(m = 0, 1, \dots)$ is also uniformly convergent on every compact subset of I.*

Proof. Let K be a compact subset of I, and we denote by u the limit function of the sequence $(u_m)_{m\in\mathbb{N}}$ on K. Then u is continuous, and $u(K)$ is a compact subset of H. If $\varepsilon > 0$ is given and x is a point in $u(K)$, then there exists a $\delta(x) > 0$ with the property that if y is an element of H with $\|y - x\| < \delta(x)$, then we have

$$\|f(t, x) - f(t, y)\| < \frac{\varepsilon}{2}\,,$$

for each t in K. If x runs through $u(K)$, then the open balls centered at x with radius $\frac{1}{2}\delta(x)$ form a covering of $u(K)$, hence, by compactness, it has a finite subcovering. The smallest of the corresponding numbers $\delta(x_i)$ we denote by δ. By the uniform convergence, there exists a natural number m_0 such that $m \geqslant m_0$ implies $\|u_m(t) - u(t)\| < \frac{1}{2}\delta$ for each t in K. On the other hand, for every t in K there exists an x_i with $\|u(t) - x_i\| \leqslant \frac{1}{2}\delta(x_i) \leqslant \frac{1}{2}\delta$, hence, for each t in K we have $\|u_m(t) - x_i\| < \delta$ for some x_i, whenever $m \geqslant m_0$. This means that for each t in K we have

$$\|f(t, u_m(t)) - f(t, u(t))\| \leqslant \|f(t, u_m(t)) - f(t, x_i)\|$$

$$+ \|f(t, u(t)) - f(t, x_i)\| < \frac{\varepsilon}{2} + \frac{\varepsilon}{2} = \varepsilon\,,$$

and the theorem is proved. □

The following lemma contains a simple but important observation.

Lemma 1.5.3. *Let I be an open interval in $\mathbb{R}$ and let t_0 be a point in I. The function $\varphi : I \to \mathbb{R}^n$ is a solution of the Cauchy problem* (1.1), (1.2) *if and only if it is continuous and satisfies*

$$\varphi(t) = x_0 + \int_{t_0}^{t} f\big(\tau, \varphi(\tau)\big) d\tau \tag{1.8}$$

for each t in I.

Proof. If φ is a solution of the Cauchy problem (1.1), (1.2), then it is continuous and we have for each t in I:

$$\varphi'(t) = f\big(t, \varphi(t)\big) . \tag{1.9}$$

Integrating this equation on the interval with endpoints t_0 and t we have

$$\varphi(t) - \varphi(t_0) = \int_{t_0}^{t} f\big(\tau, \varphi(\tau)\big) d\tau . \tag{1.10}$$

As $\varphi(t_0) = x_0$, we have (1.8).

Conversely, if $\varphi : I \to \mathbb{R}^n$ is a continuous function satisfying (1.8), then φ is differentiable, and, by differentiating (1.8), we obtain (1.1). On the other hand, (1.8) obviously implies $\varphi(t_0) = x_0$. The lemma is proved. $\square$

The following one is the most famous existence theorem in the theory of differential equations.

Theorem 1.5.1. *(Peano) If f is continuous in* (1.1), *then the Cauchy problem* (1.1), (1.2) *has a solution.*

Proof. First, we prove the following statement: if (t_0, x_0) is an arbitrary element of Ω, $f : \Omega \to \mathbb{R}^n$ is a continuous function, $\varepsilon > 0$ is arbitrary, J is a compact interval centered at t_0, and S is the closed ball centered at x_0 with radius $r > 0$, for which $J \times S$ is a subset of Ω, furthermore $M \geqslant 1$ is a bound for f on the compact set $J \times S$, then on each compact subinterval of J, whose length is at most $\frac{r}{M+\varepsilon}$ and having t_0 as the left (right) endpoint, the Cauchy problem (1.1), (1.2) has an ε-solution with values in S.

As f is uniformly continuous on $J \times S$, there is a $\delta > 0$ such that $\delta < r$, and if t_1, t_2 are in J, x_1, x_2 are in S, further $|t_1 - t_2| < \delta$, $\|x_1 - x_2\| < \delta$, then

$$\|f(t_1, x_1) - f(t_2, x_2)\| < \varepsilon .$$

Let K be a compact subinterval of J, whose length is at most $\frac{r}{M+\varepsilon}$, and whose left endpoint is t_0. We divide K with the points $t_0 < t_1 < \cdots < t_N$

into subintervals with length less than h, where $h = \min(\frac{\delta}{M}, \frac{r}{M})$. Let for $t_0 \leqslant t \leqslant t_1$

$$\varphi_\varepsilon(t) = x_0 + f(t_0, x_0)(t - t_0),$$

further let for $t_k \leqslant t \leqslant t_{k+1}$, $k = 1, 2, \ldots, N-1$

$$\varphi_\varepsilon(t) = \varphi_\varepsilon(t_k) + f(t_k, \varphi_\varepsilon(t_k))(t - t_k).$$

In order to justify this definition, we will show that $\varphi_\varepsilon(t_k)$ belongs to S for $k = 1, 2, \ldots, N-1$. Indeed, for $k = 1$ we have

$$\|\varphi_\varepsilon(t_1) - x_0\| = \|f(t_0, x_0)\|, |t_1 - t_0| < Mh \leqslant r,$$

and if we have proved this for the values $j = 1, 2, \ldots, k-1$, then for $\varphi_\varepsilon(t_k)$ we have the statement by Lemma 1.5.1. In fact, for each point t of the interval $[t_0, t_k]$ we obviously have

$$\|{\varphi_\varepsilon}'_+(t)\| \leqslant M,$$

hence

$$\|\varphi_\varepsilon(t_k) - x_0\| = \|\varphi_\varepsilon(t_k) - \varphi_\varepsilon(t_0)\| \leqslant M|t_k - t_0| \leqslant \frac{Mr}{M + \varepsilon} < r.$$

A similar argument shows that for each t in K the value $\varphi_\varepsilon(t)$ belongs to S. Obviously φ_ε is continuous on K and $\varphi_\varepsilon(t_0) = x_0$. Moreover, if $t_k < t < t_{k+1}$, then φ_ε is differentiable at t and $\varphi'_\varepsilon(t) = f\big(t_k, \varphi_\varepsilon(t_k)\big)$ $(k = 0, 1, \ldots, N-1)$. This means that $|t - t_k| < h \leqslant \delta$, and

$$\|\varphi_\varepsilon(t) - \varphi_\varepsilon(t_k)\| = \|f\big(t_k, \varphi_\varepsilon(t_k)\big)\| \, |t - t_k| < Mh \leqslant M\frac{\delta}{M} = \delta$$

implies

$$\|{\varphi_\varepsilon}'(t) - f\big(t, \varphi_\varepsilon(t)\big)\| = \|f\big(t_k, \varphi_\varepsilon(t_k)\big) - f\big(t, \varphi_\varepsilon(t)\big)\| < \varepsilon,$$

therefore the function φ_ε is an ε-solution of the Cauchy problem (1.1), (1.2) on the interval K.

A similar argument can be used to prove the statement for the intervals with right endpoint t_0. This means that if K is any compact interval in J whose length is at most $\frac{2r}{M+\varepsilon}$ with midpoint t_0, then the Cauchy problem (1.1), (1.2) has an ε-solution on K with values in S.

Now, we show that if K is a compact subinterval in J, whose length is less than $\frac{2r}{M}$ and whose midpoint is t_0, then the Cauchy problem (1.1), (1.2) has a solution on K with values in S.

Indeed, let K be a compact subinterval of J whose length is $q < \frac{2r}{M}$. Then there exists a natural number m_0 such that for each $m \geqslant m_0$ we have $q < \frac{2r}{M+\frac{1}{m}}$, hence, by the previous statements, the Cauchy problem (1.1), (1.2) has a $\frac{1}{m}$-solution on K. Let φ_m denote one of them, then we have for each t in K and $m \geqslant m_0$

$$\|\varphi_{m}{}'_{+}(t)\| \leqslant M\,,$$

therefore, by Lemma 1.5.1, it follows

$$\|\varphi_m(t) - \varphi_m(s)\| \leqslant M|t-s|$$

for all t, s in K and $m \geqslant m_0$. For $s = t_0$, we have the uniform boundedness of the sequence of functions $(\varphi_m)_{m\in\mathbb{N}}$, and for arbitrary t, s in K we have that this sequence of functions is equicontinuous for $m \geqslant m_0$. By Theorem 1.3.2 of Arzelà and Ascoli, this sequence has a subsequence $(\varphi_{m_k})_{k\in\mathbb{N}}$ which is uniformly convergent on K. Let φ denote the limit function of this sequence, then φ is obviously continuous. On the other hand, by Lemma 1.5.2, the sequence of functions $t \mapsto f\big(t, \varphi_{m_k}(t)\big)$ converges uniformly on K to the function defined by $t \mapsto f(t, \varphi(t))$. We apply Lemma 1.5.1: as we have for each t in K

$$\|\varphi_{m_k}{}'_{+}(t) - f\big(t, \varphi_{m_k}(t)\big)\| < \frac{1}{m_k}\,,$$

hence

$$\|\varphi_{m_k}(t) - x_0 - \int_{t_0}^{t} f\big(\tau, \varphi_{m_k}(\tau)\big)d\tau\| \leqslant \frac{|t-t_0|}{m_k} \leqslant \frac{2r}{Mm_k}$$

follows. If $k \to \infty$, then this implies

$$\varphi(t) = x_0 + \int_{t_0}^{t} f\big(\tau, \varphi(\tau)\big)d\tau\,,$$

whenever t is in K. This means that for any open subinterval I of J with midpoint t_0 and with length at most $\frac{2r}{M}$ we have a continuous function $\varphi : I \to S$ satisfying

$$\varphi(t) = x_0 + \int_{t_0}^{t} f\big(\tau, \varphi(\tau)\big)d\tau$$

for each t in I. This shows, by Lemma 1.5.3, that φ is a solution of the Cauchy problem (1.1), (1.2). □

In the proof of this theorem, one can recognize the so-called *Euler method* for solving first order differential equations. In fact, the ε-solution constructed in the proof which is a piecewise linear function, is an approximate solution of the Cauchy problem with error ε.

1.6 Problems

1. Give an example for a differential equation with non-continuous right side which has a unique complete solution.
2. Give an example for a Cauchy problem with continuous right side, whose solution is not locally unique.

1.7 Existence and uniqueness

In order to formulate the most effective existence and uniqueness theorem for first order differential equations we need to introduce some concepts.

Let Ω be a subset of $\mathbb{R} \times \mathbb{R}^n$, and let $f : \Omega \to \mathbb{R}^n$ be a function. We say that f satisfies the *local Lipschitz condition*, if every point (t_0, x_0) in Ω has a neighborhood $U \subseteq \Omega$, and there exists a non-negative number k such that for each (t, x_1) and (t, x_2) in U we have

$$\|f(t, x_1) - f(t, x_2)\| \leqslant k\|x_1 - x_2\| .$$

If this inequality holds for every $(t, x_1), (t, x_2)$ in Ω, then we say that f satisfies the *Lipschitz condition* with *Lipschitz constant* k. For instance, if f is continuous and it has partial derivative $\partial_2 f$ with respect to the second variable which is bounded by the constant k, then, by Lemma 1.5.1, f satisfies the Lipschitz condition with Lipschitz constant k. If f and $\partial_2 f$ are continuous, then f satisfies the local Lipschitz condition. If Ω_0 is a non-empty subset of Ω and the restriction of f to Ω_0 satisfies the (local) Lipschitz condition, then we say that satisfies the f *(local) Lipschitz condition on* Ω_0. It is easy to see that if f satisfies the local Lipschitz condition, then it satisfies the Lipschitz condition on any subset Ω_0 of Ω having compact closure $\Omega_0^{cl} \subseteq \Omega$. Here, and everywhere A^{cl} denotes the closure of the set A. Indeed, for each element (t, x) of Ω_0^{cl} let $U(t, x) \subseteq \Omega$ be a neighborhood of (t, x) with the property that f satisfies the Lipschitz condition on $U(t, x)$ with the Lipschitz constant $k(t, x)$. If (t, x) runs through Ω_0^{cl}, then the neighborhoods $U(t, x)$ form an open covering of Ω_0^{cl}, and, by its compactness, this has a finite subcovering. The maximum of the corresponding numbers $k(t, x)$ we denote by k; it is obvious that f satisfies the Lipschitz condition on Ω_0 with the Lipschitz constant k.

We say that f satisfies the *Lipschitz condition with the Lipschitz function* k, if there exists an open interval I in $\mathbb{R}$ such that for each (t, x) in Ω, t

belongs to I, $k : I \to \mathbb{R}$ is a continuous function, moreover we have for all $(t, x_1), (t, x_2)$ in Ω

$$\|f(t,x_1) - f(t,x_2)\| \leqslant k(t)\|x_1 - x_2\| \, .$$

It is obvious that if f satisfies the Lipschitz condition with some Lipschitz function, then it satisfies the local Lipschitz condition.

Theorem 1.7.1. *(Cauchy–Picard–Lindelöf) If f is continuous in* (1.1) *and it satisfies the local Lipschitz condition, then the Cauchy problem* (1.1), (1.2) *has a locally unique solution.*

Proof. Let (t_0, x_0) be an arbitrary element of Ω and for any $q > 0$, $a > 0$ we let

$$J_q =]t_0 - q, t_0 + q[, \qquad B_a = \{x : \|x - x_0\| < a\} \, ,$$

further we suppose that the set $J_q^{cl} \times B_a^{cl}$ is included in Ω. Let $\|f(t,x)\| \leqslant M$ for each t in J_q and x in B_a, further we denote by L a Lipschitz constant of f on the set $J_q^{cl} \times B_a^{cl}$. If $0 < r < q$, then F_r denotes the set of all continuous functions $y : [t_0 - r, t_0 + r] \to B_a$. Equipped with the metric

$$\rho(y,z) = \sup_{|t-t_0| \leqslant r} |y(t) - z(t)|$$

the set F_r is a complete metric space. We show that we can choose r in such a way that the following two conditions are satisfied:

i) If φ belongs to F_r, then the function $A\varphi$, defined for all t in J_r^{cl} by the equation

$$A\varphi(t) = x_0 + \int_{t_0}^{t} f\big(\tau, \varphi(\tau)\big) d\tau \, ,$$

belongs to F_r, too.

ii) There exists a number $0 < \alpha < 1$ such that if φ, ψ belong to F_r, then we have

$$\rho(A\varphi, A\psi) \leqslant \alpha \rho(\varphi, \psi) \, .$$

These two properties imply that $A : F_r \to F_r$ is a contraction.

In order that the first condition is satisfied it is necessary and sufficient that for each t in J_r^{cl} we have

$$\|A\varphi(t) - x_0\| \leqslant a \, ,$$

that is, by

$$\|A\varphi(t) - x_0\| = \|\int_{t_0}^{t} f\big(\tau, \varphi(\tau)\big)d\tau\| \leqslant M|t - t_0| \leqslant Mr\,,$$

it is enough to have $r \leqslant \frac{a}{M}$.

For the second condition, it is sufficient that for each t in J_r^{cl} we have

$$\|A\varphi(t) - A\psi(t)\| = \|\int_{t_0}^{t} \big(f\big(\tau, \varphi(\tau)\big) - f\big(\tau, \psi(\tau)\big)\big)d\tau\| \leqslant \alpha\rho(\varphi, \psi)$$

that is, by

$$\|A\varphi(t) - A\psi(t)\| = \|\int_{t_0}^{t} \big(f\big(\tau, \varphi(\tau)\big) - f\big(\tau, \psi(\tau)\big)\big)d\tau\|$$

$$\leqslant L\int_{t_0}^{t} \|\varphi(\tau) - \psi(\tau)\|d\tau \leqslant L|t - t_0|\rho(\varphi, \psi) \leqslant Lr\rho(\varphi, \psi)\,,$$

it is enough to have $r < \frac{1}{L}$. Hence if we choose $r > 0$ satisfying these two conditions, then $A : F_r \to F_r$ is a contraction, and, by Banach's Fixed Point Theorem, there exists a function φ in F_r such that $A\varphi = \varphi$, that is,

$$\varphi(t) = x_0 + \int_{t_0}^{t} f\big(\tau, \varphi(\tau)\big)d\tau$$

holds, whenever $|t - t_0| \leqslant r$. This shows that φ is a solution of the Cauchy problem (1.1), (1.2) on the interior of J_r.

To prove the local uniqueness we suppose that φ, ψ are solutions of the Cauchy problem (1.1), (1.2) on some interval I, where t_0 is in I. First, we show that if φ and ψ coincide at some point t_1 in I, then they coincide in a neighborhood of t_1. Let $x_1 = \varphi(t_1) = \psi(t_1)$. Now we consider the Cauchy problem with (t_1, x_1) instead of (t_0, x_0), and we keep our previous notation. The mapping A defined above satisfies $\varphi = A\varphi$, $\psi = A\psi$. Suppose that the number r satisfies the conditions $r \leqslant q$, $r \leqslant \frac{a}{M}$ and $r < \frac{1}{L}$, and, in addition, we assume that J_r^{cl} is a subset of I, furthermore $|t - t_1| \leqslant r$ implies $\|\varphi(t) - x_1\| \leqslant a$ and $\|\psi(t) - x_0\| \leqslant a$. This is possible, by the continuity of φ and ψ. Then φ and ψ belong to F_r, and we have

$$\rho(\varphi, \psi) = \rho(A\varphi, A\psi) \leqslant \alpha\rho(\varphi, \psi)\,,$$

which implies, by $\alpha < 1$, that $\rho(\varphi, \psi) = 0$, hence $\varphi(t) = \psi(t)$ for $|t - t_1| \leqslant r$.

Suppose now that there exists a point s in I such that $\varphi(s) \neq \psi(s)$. Let, for instance, $s > t_0$. We denote by N the set of all points in $[t_0, s]$, where

φ coincides with ψ. Obviously, t_0 belongs to N, furthermore N is a closed set. If $t_1 = \sup N$, then $t_1 < s$, however, $\varphi(t_1) = \psi(t_1)$ implies that φ and ψ coincide in a neighborhood of t_1, and this contradicts the definition of t_1. The theorem is proved. □

By the proof of Banach's Fixed Point Theorem, under the conditions of the previous theorem the solution of the Cauchy problem (1.1), (1.2) can be obtained as the uniform limit of the sequence of functions defined recursively by

$$\begin{aligned} \varphi_0(t) &= x_0 \\ \varphi_n(t) &= x_0 + \int_{t_0}^{t} f\big(\tau, \varphi_{n-1}(\tau)\big) d\tau \qquad (n = 1, 2, \dots). \end{aligned}$$

This is called the *Picard sequence* of the given Cauchy problem. The Picard sequence offers a possibility to find the solution approximately. This is called the method of *successive approximation.*

In the case of linear equations, we have the stronger existence and uniqueness result.

Theorem 1.7.2. *If* (1.1) *is linear on* $I \times \mathbb{R}^n$, *then the Cauchy problem* (1.1), (1.2) *has a unique solution on* I.

Proof. It is enough to show that the corresponding Picard sequence is uniformly convergent on any compact subinterval of I.

First of all, we note that f satisfies the Lipschitz condition with the Lipschitz function $t \mapsto \|A(t)\|$, hence it satisfies the local Lipschitz condition, hence Theorem 1.7.1 can be applied. Let $[r_1, r_2]$ be a compact subinterval of I, where $r_1 \leqslant t_0 \leqslant r_2$, and suppose that $\|A(t)\| \leqslant L$, whenever $r_1 \leqslant t \leqslant r_2$. Let $(\varphi_m)_{m \in \mathbb{N}}$ denote the corresponding Picard sequence. If C is a bound for the continuous function $\varphi_1 - \varphi_0$ on the interval $[r_1, r_2]$, and $r_1 \leqslant t \leqslant r_2$, then we have

$$\|\varphi_2(t) - \varphi_1(t)\| \leqslant \Big| \int_{t_0}^{t} \|f\big(\tau, \varphi_1(\tau)\big) - f\big(\tau, \varphi_0(\tau)\big)\| d\tau \Big| \leqslant LC|t - t_0| .$$

Similarly,

$$\|\varphi_{m+1}(t) - \varphi_m(t)\| \leqslant \frac{L^m C}{m!} |t - t_0|^m .$$

This implies for $r_1 \leqslant t \leqslant r_2$

$$\|\varphi_{m+1}(t) - \varphi_m(t)\| \leqslant C \frac{[L(r_2 - r_1)]^m}{m!} .$$

As the series $\sum \frac{[L(r_2 - r_1)]^m}{m!}$ is convergent, the series of functions

$$\varphi_0 + \sum(\varphi_{m+1} - \varphi_m)$$

is uniformly convergent on the interval $[r_1, r_2]$. On the other hand, the m-th partial sum of this series is equal to φ_{m+1}, and the theorem is proved. □

Theorem 1.7.3. *If in* (1.1) f *is continuous and satisfies the local Lipschitz condition, then the Cauchy problem* (1.1), (1.2) *has a unique complete solution which is the extension of every solution.*

Proof. We denote by Φ the set of all solutions of the Cauchy problem (1.1), (1.2), and let I be the union of the domains of the elements of Φ. Obviously, I is an open interval in $\mathbb{R}$ including t_0. We define the function $\varphi : I \to \mathbb{R}^n$ in the following way: if t is in I, then there exists a ψ in Φ such that ψ is defined at t. In this case, we let $\varphi(t) = \psi(t)$. By Theorem 1.7.1, if another solution is also defined at t, then it coincides with ψ in a neighborhood of t, hence our definition for φ is unique. Obviously, φ is the unique complete solution of the Cauchy problem (1.1), (1.2), and any solution is a restriction of φ. □

By Theorem 1.7.2, in the linear case the complete solution is defined on the whole interval I. However, a natural question arises: what is the domain of the unique complete solution in the nonlinear case? Obviously, this depends heavily on the shape of Ω, hence there is no hope to describe it in general. Nevertheless, the complete solution has a "sufficiently large" domain in the following sense.

We say that a solution φ of the Cauchy problem (1.1), (1.2) defined on I *extends to the boundary* in Ω, if for each compact subset K of Ω there exists a closed subinterval J of I such that if a point t of I does not belong to J, then the point $(t, \varphi(t))$ does not belong to K. Geometrically, this means that the curve of the solution is not included in any compact subset of the domain of the differential equation.

Theorem 1.7.4. *If in* (1.1) f *is continuous and satisfies the local Lipschitz condition, then the complete solution of the Cauchy problem* (1.1), (1.2) *extends to the boundary of* Ω.

Proof. Let K be a compact subset of Ω and let φ be the complete solution of the Cauchy problem (1.1), (1.2) with domain I. Let $m_1 = \inf I$, $m_2 = \sup I$.

We show that there are numbers r_1 and r_2 such that $m_1 < r_1 < r_2 < m_2$, further if $m_1 < t < r_1$, or $r_2 < t < m_2$, then the point $(t, \varphi(t))$ does not belong to K. We show the existence of r_2 only.

If $m_2 = +\infty$, then the statement is obvious, by the compactness of K. Let $m_2 < +\infty$ and we denote

$$\rho = \operatorname{dist}(K, \mathbb{R}^n \backslash \Omega).$$

As K is compact and $\mathbb{R}^n \backslash \Omega$ is closed, we have $\rho > 0$. Let K^* be the set of all points with distance to K not greater than $\frac{1}{2}\rho$. Obviously, K^* is a compact subset of Ω. For each (t, x) in K^*, let $U(t, x)$ be a neighborhood of (t, x) in Ω with the property that f satisfies the Lipschitz condition on $U(t, x)$ with the Lipschitz constant $k(t, x)$. If (t, x) runs through K^*, then the neighborhoods $U(t, x)$ form an open covering of K^*, hence, by the compactness, there are finitely many of them covering K^*, too. We denote by L the maximum of the corresponding Lipschitz constants, and by M a bound for f on K^*. Let further $q, a > 0$ be numbers for which $q^2 + a^2 < \frac{\rho^2}{4}$ holds. We may suppose that (t_0, x_0) belongs to K. Then for $|t - t_0| \leqslant q$ and $\|x - x_0\| \leqslant a$ the point (t, x) belongs to K^*, and if the number $r > 0$ satisfies the conditions $r \leqslant \frac{a}{M}$, $r < \frac{1}{L}$ and $r < m_2$ in Theorem 1.7.1, then φ is defined on the interval $]t_0 - r, t_0 + r[$. Let $r_2 = m_2 - r$. Suppose that $(t_1, \varphi(t_1))$ belongs to K for some $t_1 > m_2 - r$. Then the complete solution of the Cauchy problem for (1.1) with the initial condition $x(t_1) = \varphi(t_1)$ is φ again which is – by our previous considerations – defined on the interval $]t_1 - r, t_1 + r[$, hence it is defined at some points greater than m_2. This contradicts the definition of m_2. The existence of the number r_1 can be proved similarly. □

1.8 Problems

1. Let $\Omega \subseteq \mathbb{R} \times \mathbb{R}^n$ and let $f : \Omega \to \mathbb{R}^n$ be a continuous function having bounded partial derivative with respect to the second variable. Show that f satisfies the Lipschitz condition.
2. Let $\Omega \subseteq \mathbb{R} \times \mathbb{R}^n$ and let $f : \Omega \to \mathbb{R}^n$ be a continuous function having continuous partial derivative with respect to the second variable. Show that f satisfies the local Lipschitz condition.
3. Give an example for a function $f : \Omega \to \mathbb{R}^n$, where $\Omega \subseteq \mathbb{R} \times \mathbb{R}^n$ is an open set, such that f and $\partial_2 f$ is continuous, and f does not satisfy the Lipschitz condition.

4. Show that if a function satisfies the Lipschitz condition with some Lipschitz function, then it satisfies the local Lipschitz condition.
5. Using the Picard sequence, give the locally unique solution of the Cauchy problem
$$x' = x^2$$
$$x(0) = 0$$
in some neighborhood of 0 in power series form.
6. Give an example for a Cauchy problem with initial condition (t_0, x_0), whose Picard sequence does not converge uniformly in any neighborhood of t_0.

1.9 Parametric differential equations

Let Ω^* be a non-void open subset in $\mathbb{R} \times \mathbb{R}^n \times \mathbb{R}^m$, and let $f^* : \Omega^* \to \mathbb{R}^n$ be a function. If (t_0, x_0, μ_0) is in Ω^*, then those elements λ in $\mathbb{R}^m$ for which (t_0, x_0, μ) belongs to Ω^* are called *admissible parameters* with respect to the point (t_0, x_0). For each admissible parameter μ, the set Ω_μ of all those points (t, x) for which (t, x, μ) belongs to Ω^* is obviously a non-empty open set. We consider the differential equation
$$x' = f^*(t, x, \mu) \tag{1.11}$$
on this set. It is clear that if for each admissible parameter μ, the function $(t, x) \mapsto f^*(t, x, \mu)$ is continuous and satisfies the local Lipschitz condition on Ω^*, then the Cauchy problem (1.11), (1.2) has a unique complete φ_μ solution. Let Γ denote the set of all points (t, μ) for which μ is an admissible parameter with respect to the point (t_0, x_0), and φ_μ is defined at t. For each (t, μ) in Γ, we let
$$\Psi(t, \mu) = \varphi_\mu(t)\,.$$
Equation (1.11) is called first order ordinary *parametric differential equation*, the system (1.11), (1.2) is called *parametric Cauchy problem*, and the function $\Psi : \Gamma \to \mathbb{R}^n$ is called the *complete solution* of the parametric Cauchy problem (1.11), (1.2).

We shall make use of the following lemma.

Lemma 1.9.1. *(Gronwall Lemma) Let $w, \varphi, \psi : [a, b] \to \mathbb{R}$ be continuous non-negative functions such that for each $a \leqslant t \leqslant b$ we have*
$$w(t) \leqslant \varphi(t) + \int_a^t \psi(s) w(s) ds\,.$$

Then for each $a \leqslant t \leqslant b$ it follows

$$w(t) \leqslant \varphi(t) + \int_a^t \varphi(s)\psi(s) \exp\left(\int_s^t \psi(\tau)d\tau\right)ds\,.$$

Proof. Let

$$y(t) = \int_a^t \psi(s)w(s)ds\,,$$

then we have for each t in $[a, b]$

$$y'(t) - \psi(t)y(t) \leqslant \varphi(t)\psi(t)\,.$$

Let $z(t) = y(t)\exp(-\int_a^t \psi(s)ds)$, whenever t is in $[a, b]$. Then, by the previous inequality, we obtain

$$z'(t) \leqslant \varphi(t)\psi(t) \exp\left(-\int_a^t \psi(s)ds\right),$$

whenever t is in $[a, b]$. As $z(a) = 0$, hence for $a \leqslant t \leqslant b$ it follows

$$z(t) = z(t) - z(a) = \int_a^t z'(s)ds \leqslant \int_a^t \varphi(s)\psi(s) \exp\left(-\int_a^s \psi(\tau)d\tau\right),$$

consequently, by the definition of y, for $a \leqslant t \leqslant b$ we have the inequality

$$y(t) \leqslant \int_a^t \varphi(s)\psi(s) \exp\left(\int_s^t \psi(\tau)d\tau\right),$$

and finally, by $w(t) \leqslant \varphi(t) + y(t)$, our statement follows. □

The following theorem is about the complete solution of the parametric Cauchy problem.

Theorem 1.9.1. *If in* (1.11) f^* *and* $\partial_2 f^*$ *is continuous, then the complete solution* Ψ *of the parametric Cauchy problem* (1.11), (1.2) *is defined and continuous on the open set* Γ.

Proof. Let (t^*, μ^*) be a point in Γ, and suppose that, for instance, $t^* \geqslant t_0$ holds. Obviously, there exists a number r_2 in I_{μ^*} such that $t^* < r_2$. Let

$$Q = \{(t, \varphi(t, \mu^*), \mu^*) : t_0 \leqslant t \leqslant r_2\}\,.$$

It is clear that Q is a compact set, as the function $t \mapsto \varphi(t, \mu^*)$ is continuous on the interval $[t_0, r_2]$. Choose the numbers $a, b > 0$ such that if we have

$$D^* = \{(t, x, \mu) : t_0 \leqslant t \leqslant r_2,\ \|x - \varphi(t, \mu^*)\| \leqslant a,\ \|\mu - \mu^*\| \leqslant b\}\,,$$

then D^* is a subset of Ω^*. Clearly, this is possible, and D^* is a compact set. By the continuity of $\partial_2 f^*$, and by the compactness of D^*, the function

f^* satisfies the Lipschitz condition on D^*. Hence there exists a number $L > 0$ such that if (t, x, μ) and (t, y, μ) are in D^*, then

$$\|f^*(t, x, \mu) - f^*(t, y, \mu)\| \leqslant L\|x - y\|,$$

moreover, by the uniform continuity of f^* on the set D^*, there exists a decreasing function $\beta_2 : \mathbb{R}_+ \to \mathbb{R}_+$, for which $\lim_{t \to 0+} \beta_2(t) = 0$ and

$$\|f(t, x, \mu) - f(t, x, \mu^*)\| \leqslant \beta_2(\|\mu - \mu^*\|)$$

holds, whenever (t, x, μ) and (t, x, μ^*) belong to D^*.

By Theorem 1.7.4, the point $\big(t, \varphi(t, \mu), \mu\big)$ leaves the compact set D^*, as $t \to \sup I_\mu$. Let $t_2 = t_2(\mu)$ denote the smallest t for which $\big(t, \varphi(t, \mu), \mu\big)$ is a boundary point of D^*. Such a t_2 obviously exists, and it satisfies $t_0 < t_2 \leqslant r_2$. We show that $t_2 = r_2$, whenever $\|\mu - \mu^*\|$ is sufficiently small.

If $t_0 \leqslant t \leqslant t_2$, then we obviously have

$$\|\varphi(t, \mu) - \varphi(t, \mu^*\| \leqslant \int_{t_0}^{t} \big(\|f(\tau, \varphi(\tau, \mu), \mu) - f(\tau, \varphi(\tau, \mu^*), \mu)\|$$

$$+ \|f(\tau, \varphi(\tau, \mu^*), \mu) - f(\tau, \varphi(\tau, \mu^*), \mu^*)\|\big) d\tau$$

$$\leqslant \int_{t_0}^{t} \big(L\|\varphi(\tau, \mu) - \varphi(\tau, \mu^*)\| + \beta_2(\|\mu - \mu^*\|)\big) d\tau .$$

With the notation $u(t) = \|\varphi(t, \mu) - \varphi(t, \mu^*)\|$, Gronwall Lemma 1.9.1 implies for $t_0 \leqslant t \leqslant t_2$

$$\|\varphi(t, \mu) - \varphi(t, \mu^*)\| \leqslant c_2 \beta_2(\|\mu - \mu^*\|) .$$

We choose $\rho_2 > 0$ such that $\rho_2 \leqslant b$ and $c_2 \beta_2(\rho_2) < a$, further we assume that $\|\mu - \mu^*\| < \rho_2$. Then we have $\|\varphi(t_2, \mu) - \varphi(t_2, \mu^*)\| < a$, and as $\big(t_2, \varphi(t_2, \mu), \mu\big)$ is a boundary point of D^*, necessarily we infer $t_2 = r_2$.

We have shown that for each $t^* \geqslant t_0$ there exists $r_2 > t^*$ and $\rho_2 > 0$ such that $t_0 \leqslant t < r_2$, and if $\|\mu - \mu^*\| < \rho_2$, then (t, μ) belongs to the set Γ, further we have

$$\|\varphi(t, \mu) - \varphi(t, \mu^*)\| \leqslant c_2 \beta_2(\|\mu - \mu^*\|) .$$

We can prove the corresponding statement for $t^* \leqslant t_0$ similarly. Consequently, we have that for each (t^*, μ^*) in Γ there is a decreasing function $\beta : \mathbb{R}_+ \to \mathbb{R}_+$ such that $\lim_{t \to 0+} \beta(t) = 0$, further there exist numbers

$r, \rho > 0$ with $|t - t^*| < r$ such that if $\|\mu - \mu^*\| < \rho$, then (t, μ) belongs to the set Γ, moreover

$$\|\varphi(t, \mu) - \varphi(t, \mu^*)\| \leqslant c\beta(\|\mu - \mu^*\|)$$

holds, where $c \geqslant 0$ is a constant. It follows that Γ is an open set.

If (t, μ) is arbitrary in Γ, then

$$\|\varphi(t, \mu) - \varphi(t^*, \mu^*)\| \leqslant \|\varphi(t, \mu) - \varphi(t, \mu^*)\| + \|\varphi(t, \mu^*) - \varphi(t^*, \mu^*)\|$$

holds. As $t \mapsto \varphi(t, \mu^*)$ is continuous, hence, by the above argument, φ is continuous at (t^*, μ^*), that is, it is continuous on Γ. The theorem is proved. □

The above proof shows that the statement remains valid if we assume only that f^* is continuous and satisfies the local Lipschitz condition.

If the complete solution $t \mapsto \varphi(t, \mu_0)$ of (1.11) corresponding to the initial condition (t_0, x_0) is defined on the interval $[r_1, r_2]$, then there is a $\rho > 0$ such that for $\|\mu - \mu_0\| \leqslant \rho$ the solution $t \mapsto \varphi(t, \mu)$ corresponding to the initial condition (t_0, x_0) is defined on $[r_1, r_2]$, too. Moreover, for each $\varepsilon > 0$ there exists a $\delta > 0$ such that $\delta < \rho$, and $r_1 \leqslant t \leqslant r_2$, further we have for $\|\mu - \mu_0\| < \delta$ the inequality

$$\|\varphi(t, \mu) - \varphi(t, \mu_0\| < \varepsilon\,.$$

Indeed, we have seen in Theorem 1.9.1 that the set Γ is open, and (r_1, μ_0), (r_2, μ_0) belong to Γ, hence there exists $\rho > 0$ with the property that for $\|\mu - \mu_0\| \leqslant \rho$ also (r_1, μ), (r_2, μ) belong to Γ, that is, the function $t \mapsto \varphi(t, \mu)$ is defined on the interval $[r_1, r_2]$, too. On the other hand, the set of all points (t, μ) satisfying $r_1 \leqslant t \leqslant r_2$ and $\|\mu - \mu_0\| \leqslant \rho$ is compact, hence φ is uniformly continuous on it which implies the second part of the statement.

Let $\Omega^* = I \times \mathbb{R}^n \times P$ in (1.11), where I is a real open interval, and P is a non-empty open set in $\mathbb{R}^m$, further let

$$f^*(t, x, \mu) = A(t, \mu)x + b(t, \mu)\,,$$

where $A : I \times P \to M(\mathbb{R}^n)$ and $b : I \times P \to \mathbb{R}^n$ are continuous functions. Then, by Theorems 1.7.2 and 1.9.1, the domain of the complete solution of the parametric Cauchy problem (1.11), (1.2) is $I \times P$.

Lemma 1.9.2. *Let p, q be positive integers, Δ a non-empty open subset in $\mathbb{R}^p \times \mathbb{R}^q$, and $g : \Delta \to \mathbb{R}$ a continuous function having continuous partial derivative with respect to the second variable. Suppose that the set*

$$\{y : (x, y) \in \Delta \text{ for some } x\}$$

is convex. Then, with the notation $\Delta^* = \{(t, u, v) : (t, u), (t, v) \in \Delta\}$, there exists a continuous function $h : \Delta^* \to \mathbb{R}^q$ such that if (t, u, v) is in Δ^*, then

$$g(t, v) - g(t, u) = \langle h(t, u, v), (v - u) \rangle .$$

Here $\langle , \rangle$ denotes the standard inner product in $\mathbb{R}^q$.

Proof. Let $0 \leqslant s \leqslant 1$, and for each u, v in $\mathbb{R}^q$ we denote

$$w(s) = u + s(v - u) .$$

Then for each (t, u, v) in Δ^* we have

$$g(t, v) - g(t, u) = g(t, w(1)) - g(t, w(0)) = \int_0^1 \frac{d}{ds} g(t, w(s)) ds .$$

On the other hand,

$$\frac{d}{ds} g(t, w(s)) = \langle \partial_2 g(t, w(s)), (v - u) \rangle ,$$

hence the function defined by

$$h(t, u, v) = \int_0^1 \partial_2 g(t, u + s(v - u)) ds$$

whenever (t, u, v) in Δ^* satisfies the requirement. □

Theorem 1.9.2. *If in (1.11) the functions f^*, $\partial_2 f^*$ and $\partial_3 f^*$ are continuous, then the complete solution of the parametric Cauchy problem (1.11), (1.2) is continuously differentiable, further the partial derivatives $\partial_1 \partial_2 \varphi$ and $\partial_2 \partial_1 \varphi$ exist, they are continuous and equal.*

Proof. First, we introduce the following notation: $\varphi = (\varphi_1, \varphi_2, \ldots, \varphi_n)$, and $\mu = (\mu_1, \mu_2, \ldots, \mu_m)$, further $f^* = (f_1^*, f_2^*, \ldots, f_n^*)$. By Theorem 1.9.1, the complete solution φ of the parametric Cauchy problem (1.11), (1.2) is unique, and it is defined on some open set Γ. Let (t^*, μ^*) be arbitrary in Γ. First, we show that the partial derivative $\partial_2 \varphi$ exists and it is continuous at the point (t^*, μ^*). In other words, the partial derivatives $\frac{\partial \varphi_i}{\partial \mu_k}$ exist and they are continuous for $i = 1, 2, \ldots, n$; $k = 1, 2, \ldots, m$. Let J be an open interval containing the points t_0, t^*, and suppose that the function $t \mapsto \varphi(t, \mu^*)$ is defined on it. Suppose that $a, b > 0$ have the property that if t is in J^{cl}, $\|x - \varphi(t, \mu^*)\| \leqslant a$ and $\|\mu - \mu^*\| \leqslant b$, then (t, x, μ) belongs to Ω^*. By the remark following Theorem 1.9.1, there exists $\rho > 0$ such that $2\rho < b$, and for $\|\mu - \mu^*\| < 2\rho$ the function $t \mapsto \varphi(t, \mu)$ is defined on J, and for each t in J we have

$$\|\varphi(t, \mu) - \varphi(t, \mu^*\| < a\,.$$

We define

$$\Delta = \{(t, x, \mu) : t \text{ is in } J,\ \|x - \varphi(t, \mu^*)\| < a,\ \|\mu - \mu^*\| < 2\rho\}\,.$$

Obviously, Δ is an open set in $\mathbb{R} \times \mathbb{R}^n \times \mathbb{R}^m$, satisfying the conditions of Lemma 1.9.2. We denote by e_k the k-th unit vector of the standard basis of $\mathbb{R}^m$ for ($k = 1, 2, \ldots, m$), and for $\|\mu - \mu^*\| < \rho$, and $|\tau| < \rho$ we let $\nu = \mu + \tau e_k$. Then, for each t in J, we have

$$\|\varphi(t, \mu) - \varphi(t, \mu^*)\| < a, \qquad \|\varphi(t, \nu) - \varphi(t, \mu^*)\| < a\,,$$

hence for each t in J the points $(t, \varphi(t, \mu), \mu)$ and $(t, \varphi(t, \nu, \nu)$ belong to Δ. By Lemma 1.9.2, we obtain

$$f_i^*(t, \varphi(t, \nu), \nu) - f_i^*(t, \varphi(t, \mu), \mu)$$

$$= \sum_{j=1}^{n} h_{i,j}(t, \varphi(t, \mu), \mu, \varphi(t, \nu), \nu)(\varphi_j(t, \nu) - \varphi(t, \mu))$$

$$+ \sum_{j=1}^{m} h_{i,n+j}(t, \varphi(t, \mu), \mu, \varphi(t, \nu), \nu)(\nu_j) - \mu_j)\,,$$

where the functions $h_{i,j}$ are continuous for $i = 1, 2, \ldots, n; j = 1, 2, \ldots m$. We introduce the notation

$$a_{i,j}(t, \mu, \tau) = h_{i,j}(t, \varphi(t, \mu)\mu, \varphi(t, \nu), \nu)\,,$$

where $i = 1, 2, \ldots, n$, $j = 1, 2, \ldots, n + m$, t is a point in J, $\|\mu - \mu^*\| < \rho$, $|\tau| < \rho$, and $\nu = \mu + \tau e_k$. Then $a_{i,j}$ is continuous. Let for each $\tau \neq 0$

$$\psi_i(t, \mu, \tau) = \frac{1}{\tau}\big(\varphi_i(t, \nu) - \varphi_i(t, \mu)\big)\,.$$

We show that the limit $\lim_{\tau \to 0} \psi_i(t, \mu, \tau)$ exists. By (1.11), it follows

$$\frac{\partial}{\partial t}\varphi_i(t, \nu) = f_i^*(t, \varphi(t, \nu), \nu),$$
$$\frac{\partial}{\partial t}\varphi_i(t, \mu) = f_i^*(t, \varphi(t, \mu), \mu)\,,$$

hence

$$\begin{aligned}\partial_1 \psi_i(t,\mu,\tau) &= \frac{1}{\tau}\big[f_i^*(t,\varphi(t,\nu),\nu) - f_i^*(t,\varphi(t,\mu),\mu)\big]\\ &= \sum_{j=1}^{n} a_{i,j}(t,\mu,\tau)\psi_j(t,\mu,\tau) + a_{i,n+k}(t,\mu,\tau)\end{aligned}$$

holds. We have that for $\|\mu-\mu^*\|<\rho$, $0<|\tau|<\rho$ the function

$$t \mapsto (\psi_1(t,\mu,\tau),\psi_2(t,\mu,\tau),\dots,\psi_n(t,\mu,\tau))$$

is a solution on J of the parametric Cauchy problem

$$\begin{aligned} x' &= A(t,\mu,\tau)x + b(t,\mu,\tau)\\ x(t_0) &= 0\,.\end{aligned}$$

Here the entries of the matrix $A(t,\mu,\tau)$ are the numbers $a_{i,j}(t,\mu,\tau)$ whenever $i=1,2,\dots,n$ $j=1,2,\dots,n$, and

$$b(t,\mu,\tau) = (a_{1,n+k}(t,\mu,\tau), a_{2,n+k}(t,\mu,\tau),\dots,a_{n,n+k}(t,\mu,\tau))\,.$$

It follows, by the remark following Theorem 1.9.1, that the complete solution $\chi=(\chi_1,\chi_2,\dots,\chi_n)$ of the above parametric Cauchy problem is defined and continuous on the open set

$$\{(t,\mu,\tau) : t\in J,\ \|\mu-\mu^*\|<\rho,\ |\tau|<\rho\}\,.$$

Consequently, for $\tau\neq 0$ we have

$$\psi_i(t,\mu,\tau) = \chi_i(t,\mu,\tau), \qquad (i=1,2,\dots,n)\,,$$

and, by the continuity of χ_i, it follows

$$\lim_{\tau\to 0}\psi_i(t,\mu,\tau) = \lim_{\tau\to 0}\chi_i(t,\mu,\tau) = \chi_i(t,\mu,0)\,,$$

that is,

$$\frac{\partial\varphi_i}{\partial\mu_k}(t,\mu_1,\mu_2,\dots,\mu_m) = \chi_i(t,\mu_1,\mu_2,\dots\mu_m,0)$$

holds, whenever t is in J, and $\|\mu-\mu^*\|<\rho$. Hence this partial derivative is continuous in a neighborhood of the point (t^*,μ^*). Furthermore, $\frac{\partial}{\partial\mu_k}\varphi_i$ has continuous partial derivative with respect to the first variable, hence there exists the partial derivative $\partial_1\partial_2\varphi$ in a neighborhood of the point (t^*,μ^*). On the other hand, we have

$$\partial_1\varphi(t,\mu) = f^*(t,\varphi(t,\mu),\mu)$$

in a neighborhood of the point (t^*,μ^*). As the right side is continuously partially differentiable with respect to μ, hence the same holds for the left side, that is, $\partial_2\partial_1\varphi$ exists and it is continuous in a neighborhood of the point (t^*,μ^*). This implies, moreover, that $\partial_1\partial_2\varphi = \partial_2\partial_1\varphi$ holds in a neighborhood of the point (t^*,μ^*). As this point was arbitrary, our proof is complete. □

The above proof shows that if, instead of the continuity of $\partial_3 f^*$, we require the continuity of the partial derivatives of f^* with respect to some particular coordinates of the parameter only, then the continuous partial differentiability of the complete solution with respect to the same coordinates follows.

1.10 Problems

Find the derivative with respect to the parameter of the complete solution of the following parametric Cauchy problems:

1. $y' = y + \mu(x + y^2)$, $y(0) = 0$, $\frac{\partial y}{\partial \mu}|_{\mu=0} = ?$,
2. $y' = 2x + \mu y^2$, $y(0) = \mu - 1$, $\frac{\partial y}{\partial \mu}|_{\mu=0} = ?$,
3. $\frac{dx}{dt} = \frac{x}{t} + \mu t e^{-x}$, $x(1) = 1$, $\frac{\partial x}{\partial \mu}|_{\mu=0} = ?$,
4. $\left.\frac{\partial x}{\partial \mu}\right|_{\mu=0} = ?$ for

$$\begin{aligned} x' &= 4ty^2, & x(0) &= 0 \\ y' &= 1 + 5\mu x, & y(0) &= 0\,, \end{aligned}$$

1.11 Characteristic function

Let f and $\partial_2 f$ in (1.1) be continuous, further for each (τ, ξ) in Ω let $\varphi_{\tau,\xi}$ denote the complete solution of the Cauchy problem

$$\begin{aligned} x' &= f(t, x) \\ x(\tau) &= \xi\,. \end{aligned}$$

Let Γ denote the set of all points (t, τ, ξ) for which (τ, ξ) belongs to the set Ω and $\varphi_{\tau,\xi}$ is defined at t. For each such (t, τ, ξ), we define

$$\Phi(t, \tau, \xi) = \varphi_{\tau,\xi}(t)\,.$$

The function $\Phi : \Gamma \to \mathbb{R}^n$ is called the *characteristic function* of the differential equation (1.1).

Theorem 1.11.1. *If f and $\partial_2 f$ in* (1.1) *are continuous, then the characteristic function Φ of* (1.1) *is defined on an open set, it is continuous, further the partial derivatives $\partial_3\Phi$, $\partial_1\partial_3\Phi$ and $\partial_3\partial_1\Phi$ exist, they are continuous, and we have*

$$\partial_1\partial_3\Phi = \partial_3\partial_1\Phi\,.$$

Proof. We show that $\Phi : \Gamma \to \mathbb{R}^n$ is the solution of a certain parametric Cauchy problem, and then we apply Theorems 1.9.1 and 1.9.2.

Let Ω^* be the set of all points (t, x, τ, ξ) in $\mathbb{R} \times \mathbb{R}^n \times \mathbb{R} \times \mathbb{R}^n$ for which $(t + \tau, x + \xi)$ belongs to Ω. Then Ω^* is a non-empty open set, and, for each point (τ, ξ) in Ω, the point $(0, 0, \tau, \xi)$ belongs to Ω^*. On the set Ω^* we define

$$f^*(t, x, \tau, \xi) = f(t + \tau, x + \xi),$$

then f^* is a continuous function defined on Ω^* with values in $\mathbb{R}^n$, moreover $\partial_2 f^*$ and $\partial_4 f^*$ exist and they are continuous. Hence the complete solution Ψ of the parametric Cauchy problem

$$\begin{aligned} x' &= f^*(t, x, \tau, \xi) \\ x(0) &= 0 \end{aligned}$$

is defined on some open set Γ^*, it is continuous, and the partial derivatives $\partial_3 \Psi$, $\partial_1 \partial_3 \Psi$ and $\partial_3 \partial_1 \Psi$ are also continuous, further we have $\partial_1 \partial_3 \Psi = \partial_3 \partial_1 \Psi$. It is easy to see that the point (t, τ, ξ) belongs to Γ if and only if $(t - \tau, \tau, \xi)$ is in Γ^*, hence Γ is open and

$$\Phi(t, \tau, \xi) = \Psi(t - \tau, \tau, \xi) + \xi.$$

Indeed, it is easy to verify that the functions given by $t \mapsto \Phi(t, \tau, \xi)$ and $t \mapsto \Psi(t - \tau, \tau, \xi) + \xi$ are complete solutions of the Cauchy problem

$$x' = f^*(t, x) \tag{1.12}$$

$$x(\tau) = \xi, \tag{1.13}$$

hence they coincide, by uniqueness. All other statements are consequences of Theorem 1.9.2. □

If the function f in (1.1) is continuously differentiable, then the continuous differentiability of the characteristic function follows similarly. This property can be expressed by saying that in case of differentiable right side the solution of the Cauchy problem is a continuously differentiable function of the initial data. We can prove in a similar way that if the right side is r-times continuously differentiable, or analytic, then the characteristic function possesses the same properties.

Now we consider the Cauchy problem (1.12) under the conditions given above. If we introduce

$$\tilde{f}(t, x, \tau, \xi) = f^*(t + \tau, x + \xi)$$

on the appropriate domain, then, by the proof of the previous theorem, $x = y + \xi$ is the complete solution of (1.12) if and only if y is the complete solution of the parametric Cauchy problem

$$y' = \widetilde{f}(t, x, \tau, \xi) \tag{1.14}$$

$$y'(0) = 0\,. \tag{1.15}$$

This means that the Cauchy problem (1.12) can be considered as a parametric Cauchy problem, hence, using the previous theorem, we can compute the derivative of the complete solution of the Cauchy problem (1.12) with respect to the initial condition.

1.12 Problems

1. Find the derivative with respect to the initial conditions of the complete solution of the following parametric Cauchy problems:
   i) $y' = y + y^2 + xy^3,\ \ y(2) = y_0,\ \ \frac{\partial y}{\partial y_0}|_{y_0=0}\,,$
   ii) $\frac{dx}{dt} = x^2 + \mu t x^3,\ \ x(0) = 1 + \mu,\ \ \frac{\partial x}{\partial \mu}|_{\mu=0}\,,$
   iii) $\left.\frac{\partial x}{\partial y_0}\right|_{x_0=3, y_0=2}$ for

   $$x' = xy + t^2, \qquad x(1) = x_0$$
   $$2y' = -y^2, \qquad y(1) = y_0\,,$$

   iv) Find $\left.\frac{\partial y}{\partial \mu}\right|_{\mu=0}$ for

   $$x' = x + y, \qquad x(0) = 1 + \mu$$
   $$y' = 2x + \mu y^2, \qquad y(0) = -2\,.$$

2. Find the characteristic function of the following differential equations on the given domain:
   i) $x' = \frac{2x}{t+1},\ \ t > -1\,,$
   ii) $xy' + y = y^2,\ \ x > 0, y > 1\,.$

Chapter 2

ELEMENTARY SOLUTION METHODS

2.1 Separable differential equations

In this chapter, we modify our previous notation by using sometimes x instead of the independent real variable t, and the unknown function in the differential equations will be denoted by any of the letters y, u, v, etc.

Let I, J be real intervals, $f : I \to \mathbb{R}$, $g : J \to \mathbb{R}$ continuous functions, and suppose that $g(y) \neq 0$ for y in J. The differential equation

$$y' = f(x)g(y) \tag{2.1}$$

will be called *separable differential equation.* We note that, by the continuity of f and $\frac{1}{g}$, the function f has an antiderivative on I, and $\frac{1}{g}$ has an antiderivative on J, and these antiderivatives will be denoted by F and G, respectively.

Theorem 2.1.1. *The function $\varphi : I \to \mathbb{R}$ is a solution of the separable differential equation* (2.1) *if and only if the function $G \circ \varphi - F$ is constant on I.*

Proof. Suppose that $\varphi : I \to \mathbb{R}$ is a solution of (2.1) on I. Then we have for each x in I

$$(G\circ\varphi)'(x) - F'(x) = G'\big(\varphi(x)\big)\varphi'(x) - f(x) = \frac{1}{g}\big(\varphi(x)\big)f(x)g\big(\varphi(x)\big) - f(x) = 0\,,$$

hence $G \circ \varphi - F$ is constant on I. Conversely, if $G \circ \varphi - F$ is constant on I, then its derivative is zero at each point x in I which means that $\varphi : I \to \mathbb{R}$ is a solution of (2.1) on I. □

Theorem 2.1.2. *The function $\varphi : I \to \mathbb{R}$ is the solution of the separable differential equation* (2.1) *with the initial condition $y(x_0) = y_0$ if and only if $G \circ \varphi = F$.*

Proof. The statement is a consequence of the previous theorem, as the function $\frac{1}{g}$ has a constant sign on each interval. $\square$

The last two theorems mean that the solutions of the differential equation (2.1) are identical with the solutions of the algebraic equation

$$G\big(y(x)\big) - F(x) = c\,, \tag{2.2}$$

where c is an arbitrary real number. The implicit function (2.2) can be considered as "the general solution" of the differential equation (2.1) in the sense that for each solution y on some interval $I_0 \subseteq I$ there exists a constant c such that (2.2) holds on I_0, and conversely, if (2.2) holds for some differentiable function y on I with a certain constant c, then there is an open subinterval $I_0 \subseteq I$ such that y is a solution of (2.1) on I_0. Hence we get the solutions of (2.1) by "expressing" the variable y from the equation $G(y) - F(x) = c$. Practically, when solving equation (2.1) we perform the following formal computations: we replace y' by $\frac{dy}{dx}$ and we "separate" the variables symbolically in the way

$$\frac{dy}{g(y)} = f(x)\,dx\,,$$

and then we integrate both sides of the resulting "equation" with respect to "its own variable":

$$\int \frac{1}{g(y)}\,dy = \int f(x)\,dx\,.$$

After computing the indefinite integrals, we obtain the above equation

$$G(y) = F(x) + c$$

with an arbitrary constant c. The solution satisfying the initial condition $y(x_0) = y_0$ can be obtained by substituting $x = x_0$, $y = y_0$ and then the value of c can be computed.

As an example, we consider the separable differential equation

$$y' = \frac{x}{y}$$

on $\mathbb{R}_+ \times \mathbb{R}_+$. We write

$$\frac{dy}{dx} = \frac{x}{y}\,,$$

and separating the variables we get

$$y\,dy = x\,dx\,.$$

Integrating both sides we obtain

$$\int y\,dy = \int x\,dx\,,$$

that is,

$$\frac{1}{2}y^2 = \frac{1}{2}x^2 + c\,,$$

hence the solutions of the differential equation are identical with the positive solutions of the equation

$$\frac{1}{2}y^2 = \frac{1}{2}x^2 + c\,,$$

or, using another constant $c_0 = 2c$, we can write

$$y^2 - x^2 = c_0\,. \tag{2.3}$$

This is an implicit form of the general solution of the original differential equation. If $c_0 \neq 0$, then this is the equation of a hyperbola in the first quadrant, and for $c_0 = 0$ the equation describes the half-line $y = x$ in the first quadrant.

If, in addition, we have an initial condition to the above differential equation, say $y(1) = 2$, then the corresponding c_0 can be obtained by substituting $x = 1$ and $y = 2$ into (2.3):

$$4 - 1 = 3 = c_0\,,$$

hence the unique solution of the Cauchy problem satisfies

$$y^2 - x^2 = 3\,.$$

2.2 Problems

1. Find the general solution of the following separable differential equations in implicit or explicit form:

   i) $xy + (x+1)y' = 0\,,$
   ii) $\sqrt{y^2+1} = xyy'\,,$
   iii) $2x^2yy' + y^2 = 2\,,$
   iv) $y' - xy^2 = 2xy\,.$

2. Find the complete solution of the Cauchy problem

$$x' = t^2e^{-x}$$
$$x(0) = \ln 2\,.$$

3. Find the complete solution of the Cauchy problem

$$y' = \frac{4-2x}{3y^2-5}$$
$$y(1) = 3\,.$$

4. Find all complete solutions of the following differential equations:

   i) $y' = \frac{3x^2+2x+1}{y-2}$,
   ii) $(\sin x)(\sin y) + (\cos y)y' = 0$,
   iii) $xy' + y^2 + y = 0$,
   iv) $y' \ln|y| + x^2 y = 0$,
   v) $(3y^3 + 3y\cos y + 1)y' + \frac{(2x+1)y}{x^2+1} = 0$,
   vi) $x^2 yy' = (y^2-1)^{3/2}$.

5. Find all complete solutions of the following differential equations with the given domain:

   i) $y' = x^2(1+y^2)$, $\{-1 \leqslant x \leqslant 1,\ -1 \leqslant y \leqslant 1\}$,
   ii) $y'(1+x^2) + xy = 0$, $\{-2 \leqslant x \leqslant 2,\ -1 \leqslant y \leqslant 1\}$,
   iii) $y' = (x-1)(y-1)(y-2)$, $\{-2 \leqslant x \leqslant 2,\ -3 \leqslant y \leqslant 3\}$,
   iv) $(y-1)^2 y' = 2x+3$, $\{-2 \leqslant x \leqslant 2,\ -2 \leqslant y \leqslant 5\}$.

6. Find the complete solutions of the following Cauchy problems:

   i)

$$y' = \frac{3x^2+2x+1}{y-2}$$
$$y(1) = 4\,.$$

   ii)

$$y' + x(y^2+y) = 0$$
$$y(2) = 1\,.$$

2.3 Differential equations of homogeneous degree

Let I, J be open intervals, and suppose that I does not contain 0, further let $f : J \to \mathbb{R}$ be a continuous function. The differential equation

$$y' = f\left(\frac{y}{x}\right) \tag{2.4}$$

is called a *differential equation of homogeneous degree.* We note that the domain Ω of this differential equation consists of all points (x, y) in $\mathbb{R}^2$ for which x belongs to I and $\frac{y}{x}$ belongs to J.

Theorem 2.3.1. *The function $\varphi : I \to \mathbb{R}$ is a solution of the differential equation* (2.4) *if and only if the function ψ, defined at the points x of the interval I by the formula*

$$\psi(x) = \frac{\varphi(x)}{x},$$

is a solution of the separable differential equation

$$u' = \frac{f(u) - u}{x}. \tag{2.5}$$

Proof. The statement follows by simple calculation. □

We practically solve equation (2.4) in the following way: we perform the "substitution" $y = ux$ which implies, by $y' = u + u'x$, the separable equation (2.5). Solving that equation for u and substituting into $y = ux$ we have the solution y. If, in addition, we have an initial condition $y(x_0) = y_0$, then this means for u the initial condition $u(x_0) = \frac{y_0}{x_0}$.

We consider the following simple example:

$$y' = \frac{y + x}{y - x}$$

on the set $\Omega = \{(x, y) |\, x < y\}$. This equation is not a separable differential equation but it is of homogeneous degree. Indeed, we can write

$$y' = \frac{y/x + 1}{y/x - 1},$$

hence, by the substitution $y = ux$, we have

$$u'x + u = \frac{u + 1}{u - 1},$$

or

$$u'x = \frac{u + 1}{u - 1} - u = \frac{u + 1 - u^2 + u}{u - 1}.$$

Separating the variables we have the equation

$$\int \frac{1 - u}{u^2 - 2u - 1}\, du = \int \frac{dx}{x}.$$

After computing the indefinite integrals, we obtain

$$\frac{-4 + \sqrt{2}}{2} \ln |u - 1 - \sqrt{2}| + \frac{2 - \sqrt{2}}{2} \ln |u - 1 + \sqrt{2}| = \ln |x| + c.$$

Using another constant $c_0 = \exp c$ and substituting $u = \frac{y}{x}$, we have the equation

$$\frac{-4+\sqrt{2}}{2}\ln\left|\frac{y}{x} - 1 - \sqrt{2}\right| + \frac{2-\sqrt{2}}{2}\ln\left|\frac{y}{x} - 1 + \sqrt{2}\right| = \ln c_0|x|\,.$$

The function y is a solution of the original differential equation if and only if there exists a positive number c_0 such that y satisfies this latter equation. Hence this equation can be considered as the implicit form of the general solution of the original equation.

Suppose that $P : \mathbb{R} \times \mathbb{R} \to \mathbb{R}$ is a continuous function, and α is a real number. The function P is called *homogeneous of degree* α, if it satisfies

$$P(\lambda x, \lambda y) = \lambda^{\alpha} P(x, y) \tag{2.6}$$

for each positive real number λ and for all x, y in $\mathbb{R}$. More generally, P does not need to be defined on the whole $\mathbb{R} \times \mathbb{R}$, it is enough to assume that if (x, y) is in its domain, then so is $(\lambda x, \lambda y)$ for each positive real number λ.

Now let $P, Q : \mathbb{R} \times \mathbb{R} \to \mathbb{R}$ be continuous functions, α a real number and suppose that P, Q are homogeneous of degree α. Then the differential equation

$$P(x, y)y' + Q(x, y) = 0 \tag{2.7}$$

is of homogeneous degree. Indeed, we have, by the property of P and Q,

$$P(x, y)y' + Q(x, y) = x^{\alpha} P\Big(1, \frac{y}{x}\Big)y' + x^{\alpha} Q\Big(1, \frac{y}{x}\Big) = 0\,,$$

hence (2.7) is equivalent to the equation

$$P\Big(1, \frac{y}{x}\Big)y' + Q\Big(1, \frac{y}{x}\Big) = 0\,.$$

A general type of first order differential equations which can be reduced to equations of homogeneous degree is of the following form:

$$y' = f\Big(\frac{ax + by + c}{Ax + By + C}\Big)\,, \tag{2.8}$$

where f is a continuous function on some open interval, and a, b, c, A, B, C are constants. Here we suppose that $A^2 + B^2 \neq 0$, that is, the denominator of the argument of f is not constant. We shall consider two cases. In the first case we assume that the system of linear equations

$$\begin{aligned} ax + by + c &= 0 \\ Ax + By + C &= 0 \end{aligned}$$

has a unique solution (x_0, y_0). In this case, we introduce the new variables $u = x - x_0$, $v = y - y_0$. We obviously have

$$y' = \frac{dy}{dx} = \frac{dy}{dv} \cdot \frac{dv}{du} \cdot \frac{du}{dx},$$

hence substitution into (2.8) gives the new differential equation

$$\frac{dv}{du} = f\Big(\frac{au + bv + c + ax_0 + by_0}{Au + Bv + C + Ax_0 + By_0}\Big) = f\Big(\frac{au + bv}{Au + Bv}\Big). \tag{2.9}$$

We see immediately that this equation is of homogeneous degree.

In the second case the system (2.9) either has no solution, or it has infinitely many solutions, which happens if and only if there is a real number λ such that $a = \lambda A$, $b = \lambda B$. In this case we have from (2.8)

$$y' = f\Big(\frac{\lambda Ax + \lambda By + \lambda Cc + c - \lambda Cc}{Ax + By + C}\Big) = f\Big(\lambda + \frac{c(1 - \lambda C)}{Ax + By + C}\Big). \tag{2.10}$$

If $B = 0$, then this equation can be solved immediately by integration, as the right side is independent of y. If $B \neq 0$, then we apply the substitution $z = Ax + By$, then $y' = \frac{1}{B}\frac{dz}{dx}$ and we infer from (2.10)

$$z' = f\Big(\lambda + \frac{d}{z + C}\Big),$$

an autonomous, hence separable differential equation. Consequently, in any case (2.8) can be reduced either to a separable equation or to an equation of homogeneous degree.

2.4 Problems

1. Find all complete solutions of the following differential equations:

   i) $x + 2y - xy' = 0$,
   ii) $x - y + (x + y)y' = 0$,
   iii) $(y^2 - 2xy) = x^2y' = 0$,
   iv) $2x^3y' = y(2x^2 - y^2)$,
   v) $y^2 + x^2y' = xyy'$,
   vi) $(x^2 + y^2)y' = 2xy$.

2. Find all complete solutions of the following differential equations:

   i) $2x - 4y + 6 + (x + y - 3)y' = 0$,
   ii) $2x + y + 1 - (4x + 2y - 3)y' = 0$,
   iii) $x - y - 1 + (y - x + 2)y' = 0$,

iv) $(x + 4y)y' = 2x + 3y - 5$,
v) $y + 2 = (2x + y - 4)y'$,
vi) $y' = 2\left(\frac{y+2}{x+y-1}\right)$,
vii) $(y' + 1)\ln\frac{y+x}{x+3} = \frac{y+x}{x+3}$.

2.5 First order linear differential equations

Let I be an open interval and let $f, g : I \to \mathbb{R}$ be continuous functions. Then the equation

$$y' + f(x)y = g(x) \tag{2.11}$$

is called *linear first order differential equation*, or simply *linear differential equation*. If g is identically zero, then (2.11) is called a *linear homogeneous* differential equation, otherwise it is called *inhomogeneous*. The homogeneous linear differential equation

$$y' + f(x)y = 0 \tag{2.12}$$

is called the homogeneous linear differential equation *corresponding to equation* (2.11).

By Theorem 1.7.2, every complete solution of (2.11) is defined on I, and for each x_0 in I and y_0 in $\mathbb{R}$ (2.11) has a unique complete solution satisfying the initial condition $y(x_0) = y_0$. The complete solutions of (2.11) are called *integrals* of (2.11).

In this section, we shall keep our notation concerning (2.11) and (2.12).

Theorem 2.5.1. *If y_1, y_2 are integrals of* (2.11), *then $y_1 - y_2$ is an integral of* (2.12).

Proof. The statement is obvious. □

Theorem 2.5.2. *Let y_p be an integral of* (2.11). *Then every integral y of* (2.11) *has the form $y = y_h + y_p$, where y_h is an integral of* (2.12).

Proof. This follows immediately from the previous theorem. □

Theorem 2.5.3. *The set of all integrals of* (2.12) *is a one-dimensional real vector space.*

Proof. Clearly, (2.12) has non-identically zero solutions. Moreover, by uniqueness, if $y \neq 0$ is an integral of (2.12), then y is never zero. Let $y_0 \neq 0$ be an integral of (2.12). Then for each integral y of (2.12) we have

$$\left(\frac{y}{y_0}\right)' = \frac{y'y_0 - yy_0'}{y_0^2} = \frac{-f(x)yy_0 + yf(x)y_0}{y_0^2} = 0\,,$$

hence $\frac{y}{y_0}$ is constant on I. The statement is proved. □

The linear space of all integrals of (2.12) is sometimes called the *general solution* of (2.12). To describe this space we need to find a non-zero solution only. This can be done easily, as (2.12) is a separable differential equation. Applying the method we explained above, a non-zero integral of (2.12) can be written in the form

$$y_0 = \exp(-F)\,, \tag{2.13}$$

where F is an antiderivative of f on I, that is $F'(x) = f(x)$ holds for each x in I. In other words, F can be computed by indefinite integration. As soon as we have y_0 we can proceed as follows: we look for a *particular solution* y_p of (2.11) in the form

$$y_p(x) = c(x) \cdot y_0(x)$$

for each x in I, where $c : I \to \mathbb{R}$ is a differentiable function. By substitution into (2.11) we see that y_p is an integral of (2.11) if and only if $c'y_0 = g$, hence, by indefinite integration, we can compute c. This method is called the *variation of the parameter*, or *variation of the constant.* Finally, we get the general solution of (2.11) in the form

$$y = Cy_0 + y_p\,, \tag{2.14}$$

where C is an arbitrary constant, and $y_p(x) = c(x)y_0(x)$ for each x in I.

We summarize our results in the following theorem.

Theorem 2.5.4. *The function $\varphi : I \to \mathbb{R}$ is an integral of the linear first order differential equation* (2.11) *if and only if there exists a real number c such that we have*

$$\varphi = \exp(-F)(c + G)\,,$$

where F is an antiderivative of f, and G is an antiderivative of $g \exp F$.

2.6 Problems

1. Find the general solution of the following linear differential equations:

   i) $xy' - 2y = 2x^4$,
   ii) $(2x+1)y' = 4x + 2y$,
   iii) $y' + y\tan x = \sec x$,
   iv) $xy + e^x = xy'$,
   v) $x^2y' + xy + 1 = 0$,
   vi) $y = x(y' - x\cos x)$,
   vii) $2x(x^2 + y) = y'$,
   viii) $(xy' - 1)\ln x = 2y$,
   ix) $xy' + (x+1)y = 3x^2e^{-x}$.

2. Find the integral of the following Cauchy problems:

   i) $y' + \left(\frac{1+x}{x}\right)y = 0,\ \ y(1) = 1$,
   ii) $xy' + \left(1 + \frac{1}{\ln x}\right)y = 0,\ \ y(3) = 1$,
   iii) $xy' + (1 + x\cot x)y = 0,\ \ y(\pi/2) = 2$,
   iv) $y' - \left(\frac{2x}{1+x^2}\right)y = 0,\ \ y(0) = 2$.

3. Find the general solution of the following linear differential equations:

   i) $y' + 3y = 1$,
   ii) $y' + \left(\frac{1}{x} - 1\right)y = -\frac{2}{x}$,
   iii) $y' + 2xy = xe^{-x^2}$,
   iv) $y' + \left(\frac{2x}{1+x^2}\right)y = \frac{e^{-x}}{1+x^2}$,
   v) $y' + (\tan x)y = \cos x$,
   vi) $(1+x)y' + 2y = \frac{\sin x}{1+x}$,
   vii) $(x-2)(x-1)y' - (4x-3)y = (x-2)^3$,
   viii) $y' + (2\sin x\cos x)y = e^{-\sin^2 x}$,
   ix) $x^2y' + 3xy = e^x$.

4. Find the integral of the following Cauchy problems:

   i) $y' + 7y = e^{3x},\ \ y(0) = 0$,
   ii) $(1+x^2)y' + 4xy = \frac{2}{1+x^2},\ \ y(0) = 1$,
   iii) $xy' + 3y = \frac{2}{x(1+x^2)},\ \ y(-1) = 0$,
   iv) $y' + (\cot x)y = \cos x,\ \ y(\pi/2) = 1$,
   v) $y' + \frac{1}{x}y = \frac{2}{x^2} + 1,\ \ y(-1) = 0$.

2.7 Bernoulli equations

Let I be a real interval, let $f, g : I \to \mathbb{R}$ be continuous functions, further let α be a real number different from 0 and 1. The differential equation

$$y' + f(x)y = g(x)y^{\alpha} \tag{2.15}$$

is called a *Bernoulli equation.* We note that in the cases $\alpha = 0$ and $\alpha = 1$, which are excluded, this equation is linear. The solution of Bernoulli equations is based on the following theorem.

Theorem 2.7.1. *The function $\varphi : I \to \mathbb{R}$ is a solution of the Bernoulli equation* (2.15) *if and only if the function ψ defined by*

$$\psi(x) = \varphi(x)^{1-\alpha}$$

is a solution on I of the linear differential equation

$$u' + (1-\alpha)f(x)u = (1-\alpha)g(x)\,. \tag{2.16}$$

We can prove this theorem by an elementary computation. This theorem says that Bernoulli equations can be reduced to linear ones directly by the substitution $u = y^{1-\alpha}$. Obviously, in particular cases problems may arise about the possibility of this substitution depending on the sign of the unknown function, and we should study these difficulties in each particular case. We note further that if an initial condition to the equation (2.15) is also given, then via the given substitution we obtain an initial condition also for the new linear equation. Then this latter can be solved by the methods presented above.

2.8 Problems

1. Find the complete solution of the following differential equations:

   i) $u' = \frac{2}{3t}u + \frac{2t}{u}\,,$
   ii) $u' = u(1 + ue^t)\,,$
   iii) $u' = -\frac{1}{t}u + \frac{1}{tu^2}\,,$
   iv) $y' + y = y^2\,,$
   v) $7xy' - 2y = -\frac{x^2}{y^6}\,,$
   vi) $x^2y' + 2y = 2e^{1/x}y^{1/2}\,,$
   vii) $(1 + x^2)y' + 2xy = \frac{1}{(1+x^2)y}\,,$
   viii) $y' - xy = x^3y^3\,,$
   ix) $y' - \frac{1+x}{3x}y = y^4\,.$

2. Find the integral of the following Cauchy problems:

   i) $y' - 2y = xy^3, \quad y(0) = 2\sqrt{2}$,
   ii) $y' - xy = xy^{3/2}, \ y(1) = 4$,
   iii) $xy' + y = x^4y^4, \ y(1) = \frac{1}{2}$,
   iv) $y' - 2y = 2y^{1/2}, \ y(0) = 1$,
   v) $y' - 4y = \frac{48x}{y^2}, \ y(0) = 1$.

2.9 Riccati equations

Let I be a real interval and let $f, g, h : I \to \mathbb{R}$ continuous functions. The differential equation

$$y' + f(x)y = g(x)y^2 + h(x) \tag{2.17}$$

is called a *Riccati equation.* Observe that in the case $g = 0$ we have a linear equation and in the case $h = 0$ the equation is of Bernoulli-type. Anyway, we do not need to exclude these cases from the forthcoming discussion.

Theorem 2.9.1. *Let φ_p be a solution of the Riccati equation* (2.17). *The function $\varphi : I \to \mathbb{R}$ is a solution of the Riccati equation* (2.17) *if and only if the function ψ defined by*

$$\psi(x) = \varphi(x) - \varphi_p(x)$$

is a solution of the Bernoulli equation

$$u' + \big(f(x) - 2g(x)\varphi_p(x)\big)u = g(x)u^2 \,. \tag{2.18}$$

The proof is an easy calculation. By this theorem, we can find every solution of a Riccati equation by solving a Bernoulli equation, if we know a single particular solution of the Riccati equation. If φ_p is a known particular solution, then, by the substitution $u = y - \varphi_p$, we obtain a Bernoulli equation for the unknown function u. However, there is no general method for finding particular solutions of Riccati equations. In general, depending on the special form of the given functions f, g, h in the equation we should try to find a particular solution in some special class of functions, like polynomials, trigonometric polynomials, exponentials, etc.

2.10 Problems

1. Find the complete solution of the following differential equations using the given particular solution:

   i) $y' = 1 + x - (1 + 2x)y + xy^2, \ y_p(x) = 1$,

ii) $y' = e^{2x} + (1 - 2e^x)y + y^2$, $y_p(x) = e^x$,
iii) $xy' = 2 - x + (2x - 2)y - xy^2$, $y_p(x) = 1$,
iv) $xy' = x^3 + (1 - 2x^2)y + xy^2$, $y_p(x) = x$.

2. Find a particular solution and the complete solution of the following differential equations:

i) $y' + 2y = y^2 e^x$,
ii) $(x+1)(y' + y^2) = -y$,
iii) $xy^2 y' = x^2 + y^3$,
iv) $xydy = (y^2 + x)dx$,
v) $xy' + 2y + x^5 y^3 e^x = 0$.

2.11 Exact differential equations

Let $D \subseteq \mathbb{R} \times \mathbb{R}$ be a non-empty open set, further let $P, Q : D \to \mathbb{R}$ be continuous functions satisfying $P(x,y)^2 + Q(x,y)^2 > 0$ at each point (x,y) in D. We agree on that the equation

$$P(x,y)dx + Q(x,y)dy = 0 \tag{2.19}$$

means the differential equation

$$y' = -\frac{P(x,y)}{Q(x,y)} \tag{2.20}$$

on each non-empty, open, and connected subset Ω of D such that for (x,y) in Ω we have $Q(x,y) \neq 0$, and it means the differential equation

$$x' = -\frac{Q(x,y)}{P(x,y)} \tag{2.21}$$

on each non-empty, open, connected subset Ω of D such that for (x,y) in Ω we have $P(x,y) \neq 0$. The following question arises: if there is a non-empty open and connected subset Ω of D such that neither of Q and P vanishes on Ω, then what is the meaning of equation (2.19)? In these cases, the right sides of equations (2.20) and (2.21) have constant sign, hence all solutions of these equations are strictly monotonic functions, moreover, it is obvious that the function φ is a solution of (2.20) on some interval I if and only if its inverse, φ^{-1} is a solution of (2.21) on the interval $\varphi(I)$. This means that in this case equation (2.19) may mean any of the equations (2.20) or (2.21), because from the solutions of any of them we can obtain easily the solutions of the other. We note that – according to this agreement – all separable equations can be written in the form

$$f(x)dx + g(y)dy = 0,$$

and the equations of homogeneous degree can be written in the symmetric form (2.19), where the functions P and Q are homogeneous functions of the same degree. In the subsequent paragraphs we always suppose concerning (2.19) that it means equation (2.20) – the other case can be treated similarly.

The differential equation (2.19) is called *exact differential equation*, if there exists a differentiable function $F : D \to \mathbb{R}$ such that $\partial_1 F = P$ and $\partial_2 F = Q$ holds. Such a function F is called a *primitive function* of the pair of functions P, Q, or of the vector function (P, Q). It plays the role of an antiderivative.

Theorem 2.11.1. *Let $D \subseteq \mathbb{R} \times \mathbb{R}$ be a non-empty open set, further let equation* (2.19) *be exact. Then the function $\varphi : I \to \mathbb{R}$ is a solution of equation* (2.19) *if and only if for some primitive function F of the pair of functions P, Q the function $x \mapsto F\big(x, \varphi(x)\big)$ is constant on I.*

Proof. If the function $\varphi : I \to \mathbb{R}$ is a solution of the exact differential equation (2.19), and F is a primitive function of the pair of functions P, Q, then, by the Chain Rule, the derivative of the function $x \mapsto F(x, \varphi(x))$ is

$$\partial_1 F(x, \varphi(x)) + \partial_2 F(x, \varphi(x))\varphi'(x) = P(x, \varphi(x)) + Q(x, \varphi(x))\varphi'(x) = 0$$

at each point x in the interval I, hence this function is constant on I. The converse statement is also obvious, because the derivative of a constant is zero, which implies that (2.19) holds for φ. □

By this theorem, the general solution of an exact equation can be given in the implicit form

$$F(x, y) = c, \tag{2.22}$$

whenever F is a primitive function of the pair (P, Q). This equation is symmetric in the variables x, y which explains the advantage of the way of writing differential equations in the form (2.21): the two variables can be treated symmetrically – it makes no difference if we determine the function itself, or its inverse.

Let $D \subseteq \mathbb{R} \times \mathbb{R}$ be a non-empty open set and let $P, Q : D \to \mathbb{R}$ be continuously differentiable functions. Suppose that $F : D \to \mathbb{R}$ is a primitive function of the pair of functions P, Q. Then we have the equations $\partial_2 P = \partial_2 \partial_1 F = \partial_1 \partial_2 F = \partial_1 Q$ on D. Hence the condition $\partial_2 P = \partial_1 Q$ on D is necessary for the equation (2.19) is exact, whenever the functions P, Q

are continuously differentiable. We know from elementary analysis, from the theory of line integrals, that under different assumptions on the set D this condition is also sufficient. For instance, if D is a non-empty open and convex set and the continuously differentiable functions $P, Q : D \to \mathbb{R}$ satisfy the equation $\partial_2 P = \partial_1 Q$ on D, then the pair of functions P, Q has an antiderivative $F : D \to \mathbb{R}$ which can be determined in the following way: let (x_0, y_0) and (x, y) be arbitrary points in D and let $g : [a, b] \to D$ be a piecewise smooth curve with $g(a) = (x_0, y_0)$, $g(b) = (x, y)$. In other words, g is a piecewise smooth curve in D which joins the points (x_0, y_0) and (x, y). For instance, by the convexity of D, the curve g can always be chosen as the line segment joining the two points. If $F(x, y)$ denotes the line integral of the function (P, Q) along the curve g, then, by the theory of line integrals, the function $(x, y) \mapsto F(x, y)$ is an antiderivative of the pair of functions P, Q. If $g = (g_1, g_2)$ is smooth, that is, continuously differentiable, then

$$F(x, y) = \int_a^b \Big[P(g_1(t), g_2(t)) g_1'(t) + Q(g_1(t), g_2(t)) g_2'(t) \Big] \, dt \,.$$

We can determine a primitive function in a similar manner on star-like, or simply connected sets D. The particular choice of the curve g depends on the particular shape of D, of course. In most cases we choose a piecewise linear curve joining the two points which consists of line segments parallel to the coordinate axes, supposing that this choice is possible, depending on the shape of D.

If an initial condition $y(x_0) = y_0$ is also given to the exact equation (2.19), where (x_0, y_0) is a point in D, further F is a primitive function of the pair of functions P, Q, then each solution $\varphi : I \to \mathbb{R}$ of the corresponding Cauchy problem satisfies the equation

$$F\big(x, \varphi(x)\big) = F(x_0, y_0)$$

for every x in I. Hence the solutions of the Cauchy problem can be computed formally by "expressing" the variable y from the equation

$$F(x, y) = F(x_0, y_0) \,.$$

Theoretically this is possible locally, by the Implicit Function Theorem, as the partial derivative $\partial_2 F = Q$ is different from zero at the points of D, however, in particular cases this can be very difficult, or even impossible to perform. Therefore the "general solution" of (2.19) is considered as the implicit equation

$$F(x, y) = c \,,$$

and the implicit equation

$$F(x, y) = F(x_0, y_0)$$

is considered as the "solution" of the corresponding Cauchy problem.

By the definition of the exact differential equation, it follows that if we multiply both sides of an equation of the form (2.19) by the same positive continuously differentiable function, then it may happen that the original equation was exact, but after the multiplication the new equation is not exact anymore, or conversely, although the two equations are equivalent in the sense that, obviously, they have the same solutions. This means that exactness is actually not a property of the differential equation, but rather of its special form. For instance, the separable equation

$$f(x)dx - \frac{dy}{g(y)} = 0$$

is obviously exact, but if we write the equation in the (equivalent) form

$$g(y)dx - \frac{dy}{f(x)} = 0\,,$$

then this can be exact only in some trivial cases. Therefore the following problem arises: given equation (2.19), under which conditions does there exist a positive and continuously differentiable function $\mu : D \to \mathbb{R}$ such that the differential equation

$$\mu(x, y)P(x, y)dx + \mu(x, y)Q(x, y)dy = 0 \tag{2.23}$$

is exact? Such a function μ is called an *integrating factor*, or *Euler multiplier*. If D is a non-empty open and convex set, and the functions μ, P, Q are continuously differentiable, then, by the above considerations, a necessary and sufficient condition for this is that $\partial_2(\mu\, P) = \partial_1(\mu\, Q)$ holds on D. In particular, this means that

$$\partial_2\mu\, P + \mu\partial_2 P = \partial_1\mu\, Q + \mu\partial_1 Q\,,$$

or, in a more suggestive form

$$\mu_y P - \mu_x Q = \mu(Q_x - P_y)\,. \tag{2.24}$$

This is a *partial differential equation* for the unknown function μ. It seems that we have "reduced" our original problem to a much more complicated one, namely, to the solution of a partial differential equation. Nevertheless, we do not need to solve this partial differential equation: we have to find only a solution μ of it which is positive and continuously differentiable.

This means that we can try to look for solutions of (2.24) having some special form: it depends only on x, only on y, or it depends only on some special function of these two variables (sum, difference, product, etc.). In these cases equation (2.24) reduces to an ordinary first order differential equation, and we need only to find a positive solution of it. For instance, if we suppose that the integrating factor μ depends only on x, then $\partial_2 \mu = 0$, hence $\mu(x,y) = \mu(x)$, and from equation (2.24) we derive the equation

$$\frac{\mu'}{\mu} = \frac{P_y - Q_x}{Q} .$$

Here the left hand side depends on x only, hence a solution μ of this type exists if and only if the right hand side depends on x only, too. Then we have

$$\mu = \exp\left(\int \frac{P_y - Q_x}{Q} \right) .$$

We can find simple necessary and sufficient conditions for the existence of integrating factors of different special form in a similar way. The details are left to the reader.

2.12 Problems

1. Determine which of the following equations are exact and find their complete solutions:

   i) $6x^2y^2dx + 4x^3ydy = 0$,
   ii) $(x^2 - y)dx - xdy = 0$,
   iii) $(3y\cos x + 4xe^x + 2x^2e^x)dx + (3\sin x + 3)dy = 0$,
   iv) $y(x - 2y)dx - x^2dy = 0$,
   v) $14x^2y^3dx + 21x^2y^2dy = 0$,
   vi) $(x^2 + y^2)dx + 2xydy = 0$,
   vii) $(2x - 2y^2)dx + (12y^2 - 4xy)dy = 0$,
   viii) $(1 + e^{2\theta})d\rho + 2\rho e^{2\theta}d\theta = 0$,
   ix) $(x + y)^2dx + (x + y)^2dy = 0$,
   x) $(2x + 3y + 4)dx + (3x + 4y + 5)dy = 0$,
   xi) $(4x + 7y)dx + (3x + 4y)dy = 0$,
   xii) $(-2y^2\sin x + 3y^3 - 2x)dx + (4y\cos x + 9xy^2)dy = 0$,
   xiii) $(2x + y)dx + (2y + 2x)dy = 0$,
   xiv) $(3x^2 + 2xy + 4y^2)dx + (x^2 + 8xy + 18y)dy = 0$.

2. Find the complete solution of the following Cauchy problems:

   i) $(4x^3y^2 - 6x^2y - 2x - 3)dx + (2x^4y - 2x^3)dy = 0,\ \ y(1) = 3\,,$
   ii) $(-4y\cos x + 4\sin x\cos x + \sec^2 x)dx + (4y - 4\sin x)dy = 0,\ \ y(\frac{\pi}{4}) = 0\,,$
   iii) $(y^3 - 1)e^x dx + 3y^2(e^x + 1)dy = 0,\ \ y(0) = 0\,,$
   iv) $(\sin x - y\sin x - 2\cos x)dx + \cos x dy = 0,\ \ y(0) = 1\,,$
   v) $(2x - 1)(y - 1)dx + (x + 2)(x - 3)dy = 0,\ \ y(1) = -1\,.$

3. Find all functions M such that the given equation is exact:

   i) $M(x,y)dx + (x^2 - y^2)dy = 0\,,$
   ii) $M(x,y)dx + 2xy\sin x\cos y dy = 0\,,$
   iii) $M(x,y)dx + (e^x - e^y\sin x)dy = 0\,.$

4. Find all functions N such that the given equation is exact:

   i) $(x^3y^2 + 2xy + 3y^2)dx + N(x,y)dy = 0\,,$
   ii) $(\ln xy + 2y\sin x)dx + N(x,y)dy = 0\,,$
   iii) $(x\sin x + y\sin y)dx + N(x,y)dy = 0\,.$

5. Find an integrating factor of only one variable and find the complete solution of the given equations:

   i) $ydx - xdy = 0\,,$
   ii) $3x^2ydx + 2x^3dy = 0\,,$
   iii) $2y^3dx + 3y^2dy = 0\,,$
   iv) $(5xy + 2y + 5)dx + 2xdy = 0\,,$
   v) $(xy + x + 2y + 1)dx + (x + 1)dy = 0\,.$

6. Find an integrating factor of the form $\mu(x,y) = p(x)\cdot q(y)$ and find the complete solution of the following equations:

   i) $y(1 + 5\ln|x|)dx + 4x\ln|x|dy = 0\,,$
   ii) $(3x^2y^3 - y^2 + y)dx + (-xy + 2x)dy = 0\,,$
   iii) $2ydx + 3(x^2 + x^2y^3)dy = 0\,,$
   iv) $x^4y^4dx + x^5y^3dy = 0\,,$
   v) $y(x\cos x + 2\sin x)dx + x(y + 1)\sin x dy = 0\,.$

2.13 Incomplete differential equations

In the applications we often have to consider non-explicit differential equations which means that the derivative of the unknown function is not explicitly expressed in terms of the independent variable x and the dependent variable y. A general form can be given as follows. Let $D \subseteq \mathbb{R}^3$ be a non-empty open set and $F\colon D \to \mathbb{R}$ a continuous function with continuous partial derivative $\partial_3 F$ which does not vanish identically on any non-empty open subset of D. Then

$$F(x, y, y') = 0 \tag{2.25}$$

is called a first order *implicit differential equation*, or simply an *implicit differential equation*. The condition on $\partial_3 F$ implies, by the Implicit Function Theorem, that y' can be expressed locally from the equation, hence, at least locally, (2.25) is equivalent to an explicit differential equation. For our purposes this condition assures that our equation is a differential equation indeed, that is, y' appears in the equation. An initial condition to (2.25) can be given in the following manner: let $D_{1,2}$ denote the set

$$D_{1,2} = \{(x, y) : \text{there exists } p \text{ such that } (x, y, p) \in D\}.$$

Obviously, $D_{1,2} \subseteq \mathbb{R}^2$ is a non-empty open set. Then the equation

$$y(x_0) = y_0 \tag{2.26}$$

is called an *initial condition* for (2.25), if (x_0, y_0) is in $D_{1,2}$. In this volume we will not go into the theory of implicit differential equations, we just highlight some special cases, where the solutions can be obtained using simple methods. Nevertheless, we call the reader's attention that our theory does not guarantee nor existence, neither uniqueness for implicit differential equations.

The implicit differential equation (2.25) will be called *incomplete*, if either of the variables x and y, or both do not appear in (2.25). We shall consider the following three types of incomplete equations:

$$x = f(y'), \tag{2.27}$$

$$F(x, y') = 0, \tag{2.28}$$

and

$$F(y, y') = 0. \tag{2.29}$$

Obviously, (2.27) is a special case of (2.28). In all these cases, we impose reasonable conditions on the functions f, F in order to avoid trivial cases, like f, F are continuous and nonidentically zero, they are differentiable sufficiently many times, and F depends properly on the second variable, etc. In this section, we show how we can find solutions of these types of equations – at least locally.

We introduce the notation $p = y'$ and we shall consider p as a parameter. Then (2.27) can be written in the form $x = f(p)$, hence $dx = f'(p)dp$. On the other hand, by $dy = pdx$, we have

$$dy = pf'(p)dp\,, \tag{2.30}$$

which is an explicit equation for y, as a function of p. Solving it we obtain the solution of (2.27) in the following parametric form

$$\begin{aligned} x &= f(p) \\ y &= g(p)\,, \end{aligned}$$

where g is the solution of (2.30).

Consider the following example:

$$e^{y'} + y' = x\,. \tag{2.31}$$

In this case, we have $x = e^p + p$, hence $dx = (e^p + 1)dp$ which implies

$$dy = pdx = p(e^p + 1)dp\,.$$

In other words, we have to solve the differential equation

$$\frac{dy}{dp} = pe^p + p\,. \tag{2.32}$$

By integration, we have $y = (p-1)e^p + \frac{p^2}{2} + C$, and the parametric form of the solution is

$$\begin{aligned} x(p) &= e^p + p \\ y(p) &= (p-1)e^p + \frac{p^2}{2} + C\,, \end{aligned}$$

where C is an arbitrary constant.

In the case of equation (2.28), we suppose that a parametric system can be obtained for x and p in the form $x = \varphi(t)$, $p = \psi(t)$. Then we have $dy = pdx = \psi(t)\varphi'(t)dt$ which is a differential equation

$$\frac{dy}{dt} = \psi(t)\varphi'(t) \tag{2.33}$$

for y as a function of t. Solving it we obtain the solution of (2.28) in the following parametric form

$$\begin{aligned} x &= \varphi(t) \\ y &= \chi(t)\,, \end{aligned}$$

where χ is the solution of (2.33).

We consider the example

$$x^3 + y'^3 = 3xy' . \tag{2.34}$$

We introduce the parameter t by $p = y' = tx$, then we obtain

$$x^3 + t^3x^3 = 3tx^2 ,$$

which gives

$$x = \frac{3t}{1+t^3}, \quad p = \frac{3t^2}{1+t^3} .$$

Now we have

$$dy = pdx = \frac{3t^2}{1+t^3}\frac{3(1-2t^3}{(1+t^3)^2}dt .$$

Finally, by integration, we obtain

$$y(t) = -\frac{9}{2(1+t^3)^2} + \frac{6}{1+t^3} + C .$$

Consequently, the parametric form of the solution is

$$\begin{aligned} x(t) &= \frac{3t}{1+t^3} \\ y(t) &= -\frac{9}{2(1+t^3)^2} + \frac{6}{1+t^3} + C , \end{aligned}$$

where C is an arbitrary constant.

In the third case of (2.29) we interchange the role of x and y and we consider the problem for the function $x = x(y)$. In this case $dx = \frac{1}{p}dy$, and we apply the above ideas.

In the case of the equation

$$y = y' + \ln y'$$

we have $dy = (1 + \frac{1}{p})dp$, and $dx = \frac{1}{p}dy = \left(\frac{1}{p} + \frac{1}{p^2}\right)dp$, hence, by integration, we infer that $x = \ln p - \frac{1}{p} + C$. The parametric solution is

$$\begin{aligned} x(p) &= \ln p - \frac{1}{p} + C \\ y(p) &= p + \ln p , \end{aligned}$$

where C is an arbitrary constant.

In the following example we consider the differential equation

$$y'^3 - y^2(1 - y') = 0 .$$

We can write $p^3 - y^2(1-p) = 0$ and we introduce the parameter t by the equation $y = tp$. Then we have

$$p = \frac{t^2}{1+t^2}, \quad y = \frac{t^3}{1+t^3}.$$

From the equation $dx = \frac{dy}{p}$ we obtain

$$\frac{dx}{dt} = \frac{3+t^2}{1+t^2},$$

and finally, by integration,

$$x = t - 2\tan^{-1} t + C.$$

Hence the parametric form of the solution is the following:

$$x(t) = t - 2\tan^{-1} t + C$$
$$y(t) = \frac{t^3}{1+t^2},$$

where C is an arbitrary constant.

2.14 Problems

Solve the following incomplete differential equations:

1. $xy'^3 = 1 + y'$,
2. $y'^3 - x^3(1-y') = 0$,
3. $y'^3 + y^3 = 3yy'$,
4. $y = e^{y'}y'^2$,
5. $y^2(y'-1) = (2-y')^2$,
6. $y(1+y'^2) = 2$.

2.15 Implicit differential equations

In some cases the general implicit differential equation (2.25) can be solved by introducing parameters, too. Geometrically, equation (2.25) can be considered as an equation of a surface in $\mathbb{R}^3$. Suppose that we have a parametric representation of this surface in the form

$$x = \varphi(u,v) \tag{2.35}$$
$$y = \psi(u,v) \tag{2.36}$$
$$p = \chi(u,v). \tag{2.37}$$

Here we use the notation $p = y'$ again. Hence we have $dy = pdx$. Further, using (2.35) we have

$$\partial_u\psi\, du + \partial_v\psi\, dv = \chi(u,v)\big[\partial_u\varphi\, du + \partial_v\varphi\, dv\big].$$

This is a first order differential equation between u and v. If we choose u as the independent variable, then we have

$$\frac{dv}{du} = \frac{\chi\partial_u\varphi - \partial_u\psi}{\partial_v\psi - \chi\partial_v\varphi}.$$

Suppose that the solution of this equation is $\omega = \omega(u, C)$, then, by substitution, we have the solution of the original equation in the parametric form

$$x = \varphi\big(u, \omega(u, C)\big), \quad y = \psi\big(u, \omega(u, C)\big),$$

where C is an arbitrary constant.

We consider the special case

$$y = f(x, y'). \tag{2.38}$$

That means, $y = f(x, p)$ and $dy = pdx$. Taking differentials we have

$$pdx = dy = \partial_x f\, dx + \partial_p f\, dp,$$

which is a first order equation between x and p. Solving it for p we have $p = \varphi(x, C)$, and $y = f\big(x, \varphi(x, C)\big)$.

In the other special case

$$x = f(y, y') \tag{2.39}$$

we interchange the role of x and y and apply the previous argument.

Now we have the following example:

$$y'^3 - 4xyy' + 8y^2 = 0.$$

Using the notation $p = y'$, we write

$$p^3 - 4xyp + 8y^2 = 0. \tag{2.40}$$

Assuming $y \neq 0$, $y' \neq 0$, we obtain

$$x = \frac{p^2}{4y} + \frac{2y}{p}.$$

We consider x, p as functions of y, and differentiate with respect to y. It follows

$$\frac{dp}{dy} \cdot \frac{p^3 - 4y^2}{2yp^2} = \frac{p^3 - 4y^2}{4y^2 p}.$$

Suppose that $p^3 \neq 4y^2$. Then we get the differential equation

$$\frac{dp}{p} = \frac{1}{2}\frac{dy}{y}$$

with the solution $p(y) = C_1 y^{1/2}$, where C_1 is an arbitrary constant. Substitution into (2.40) gives the solution

$$y(x) = C(x - C)^2\,,$$

where $C = \frac{C_1^2}{4}$. On the other hand, if $p^3 = y'^3 = 4y^2$, then this is a separable differential equation for y which gives another set of solutions.

2.16 Problems

Solve the following implicit differential equations:

1. $x = y'^3 + y'\,,$
2. $x(y'^2 - 1) = 2y'\,,$
3. $x = y'\sqrt{y'^2 + 1}\,,$
4. $y'(x - \ln y') = 1\,,$
5. $y = y'^2 + 2y'^3\,,$
6. $y = \ln(1 + y'^2)\,,$
7. $(y' + 1)^3 = (y' - y)^2\,,$
8. $y = (y' - 1)e^{y'}\,,$
9. $y'^4 - y'^2 = y^2\,,$
10. $y'^2 - y'^3 = y^2\,,$
11. $y'^4 = 2yy' + y^2\,,$
12. $y'^2 - 2xy' = x^2 - 4y\,,$
13. $5y + y'^2 = x(x + y')\,,$
14. $x^2y'^2 = xyy' + 1\,,$
15. $y'^3 + y^2 = xyy'\,,$
16. $2xy' - y = y' \ln yy'\,,$
17. $y' = e^{xy'/y}\,,$
18. $y = xy' - x^2y'^3\,.$

2.17 Lagrange and Clairut equations

The differential equation

$$A(y')y + B(y')x = C(y') \tag{2.41}$$

is called *Lagrange equation*, if $A, B, C : I \to \mathbb{R}$ are continuously differentiable functions on the non-empty open interval $I \subseteq \mathbb{R}$. Using the notation

$p = y'$ we may suppose that $A \neq 0$, and then we express y from the equation in the form

$$y = \varphi(p)x + \psi(p)\,. \tag{2.42}$$

Here $\varphi, \psi : I \to \mathbb{R}$ are continuously differentiable. We apply the method of differentiation used in the previous section – this time we consider y, p as functions of x and we differentiate with respect to x. It follows

$$p = \varphi(p) + \big[\varphi'(p)x + \psi'(p)\big]\frac{dp}{dx}\,.$$

This is a linear equation for x:

$$\frac{dx}{dp} + \frac{\varphi'(p)}{\varphi(p) - p}x = \frac{\psi'(p)}{p - \varphi(p)}\,.$$

Here we have to assume that $\varphi(p) \neq p$. The solution can be obtained in the form

$$x(p) = C\omega(p) + \chi(p)\,.$$

Substituting into (2.42) we have the solution of (2.41) in the parametric form

$$\begin{aligned} x(p) &= C\omega(p) + \chi(p) \\ y(p) &= \big[C\omega(p) + \chi(p)\big]\varphi(p) + \psi(p)\,. \end{aligned}$$

Concerning the assumption $\varphi(p) \neq p$ we note that if there is a constant C_0 satisfying $\varphi(C_0) = C_0$, then we obviously have the solution $y = \varphi(C_0)x + \psi(C_0)$ which is not included in the family of solutions given above. Such solutions are called *singular solutions*. The study of singular solutions is out of the scope of this volume.

A special case of Lagrange equation (2.41) is the differential equation which is called *Clairut equation*:

$$y = xy' + \varphi(y') = xp + \varphi(p)\,. \tag{2.43}$$

Applying the above method we arrive at the equation

$$\frac{dp}{dx}\big[x + \varphi'(p)\big] = 0\,.$$

If $\frac{dp}{dx} = 0$, then we have $p = C$, and substitution gives $y = Cx + \varphi(C)$, where C is an arbitrary constant. This is a one parameter family of straight lines. Otherwise we have $x + \varphi'(p) = 0$, that is, $x = -\varphi'(p)$. This equation determines p as the function of x in the form $p = \omega(x)$. Substitution gives

$$y = x\omega(x) + \varphi\big(\omega(x)\big)\,.$$

Here we substitute $x = -\varphi'(p)$ to get the solution in parametric form:

$$\begin{aligned} x(p) &= -\varphi'(p) \\ y(p) &= -p\varphi'(p) + \varphi(p)\,. \end{aligned}$$

2.18 Problems

Solve the following Lagrange and Clairut differential equations:

1. $y = xy' - y'^2$,
2. $y + xy' = 4\sqrt{y'}$,
3. $y = 2xy' - 4y'^3$,
4. $y = xy' - (2 + y')$,
5. $y'^3 = 3(xy' - y)$,
6. $y = xy'^2 - 2y'^3$,
7. $xy' - y = \ln y'$,
8. $xy'(y' + 2) = y$,
9. $2y'^2(y - xy') = 1$,
10. $2xy' - y = \ln y'$.

Chapter 3

LINEAR DIFFERENTIAL EQUATIONS

3.1 Integrals of linear differential equations

In Section 2.5 we considered first order linear differential equations. In this chapter we generalize the results obtained there for higher dimension, that is, for the case $n > 1$ in (1.3):

$$x' = A(t)x + b(t)\,, \tag{1.3}$$

During the forthcoming study of linear differential equations we shall include complex valued solutions, too. Obviously, in this case we allow complex valued functions A and b also on the right hand side of (1.3), that is, $A : I \to M(\mathbb{C}^n)$ and $b : I \to \mathbb{C}^n$ are continuous functions. Clearly, a differentiable function $\varphi : I \to \mathbb{C}^n$ is a solution of (1.3) if and only if substituting $x = \varphi$ into (1.3) the real and imaginary parts of the two sides of the equation are equal. Obviously, in this case x_0 appearing in the initial condition (1.2):

$$x(t_0) = x_0 \tag{1.2}$$

can be an arbitrary element of $\mathbb{C}^n$. It is easy to see that the theorems verified above for linear differential equations remain valid in this case. We shall use this fact in the future. We also use the notation $\mathbb{K}$ for $\mathbb{R}$ or $\mathbb{C}$.

Suppose that $A : I \to M(\mathbb{K}^n)$ and $b : I \to \mathbb{K}^n$ are continuous functions in (1.3). Every complete solution of (1.3) – which is defined on the whole interval I, by Theorem 1.7.2, – will be called an *integral* of (1.3). We recall that (1.3) is homogeneous, if $b = 0$, otherwise it is called inhomogeneous, and (1.4):

$$x' = A(t)x \tag{1.4}$$

is the homogeneous equation corresponding to (1.3). For each t in I and x in $\mathbb{K}^n$ let $\varphi_{t,x}$ denote the integral of (1.4) satisfying the initial condition $\varphi_{t,x}(t) = x$. Then $\varphi_{t,x}$ is defined on the interval I, and the function defined by $(s,t,x) \mapsto \varphi_{t,x}(s)$ is continuous on $I \times I \times \mathbb{K}^n$. For each s, t in I and x in $\mathbb{K}^n$ we let

$$\Phi(s,t,x) = \varphi_{t,x}(s).$$

We realize that the function $\Phi : I \times I \times \mathbb{K}^n \to \mathbb{K}^n$ is the characteristic function of (1.4). Further, for each s, t in I and x in $\mathbb{K}^n$ we let

$$R(s,t)(x) = \Phi(s,t,x).$$

The function $R : I \times I \to \mathbb{K}^n$ is called the *resolvent* of (1.4).

Theorem 3.1.1. *If R is the resolvent of* (1.4)*, then $R(s,t)$ is an automorphism of the linear space $\mathbb{K}^n$, further we have*

$$R(s,t) \cdot R(t,u) = R(s,u) \tag{3.1}$$

for each s, t, u in I.

Proof. The linearity of $R(s,t)$ follows from the fact that for any fixed x, y in $\mathbb{K}^n$ and λ, μ in $\mathbb{K}$ both functions

$$s \mapsto \Phi(s,t,\lambda x + \mu y)$$

and

$$s \mapsto \lambda\Phi(s,t,x) + \mu\Phi(s,t,y)$$

are solutions of (1.4) on I with the initial condition $(t, \lambda x + \mu y)$, hence they are equal, by uniqueness.

For any fixed x in $\mathbb{K}^n$ and t, u in I both functions

$$s \mapsto \Phi(s,u,x)$$

and

$$s \mapsto \Phi\big(s,t,\Phi(t,u,x)\big)$$

are solutions of (1.4) on I with the initial condition $\big(t, \Phi(t,u,x)\big)$, hence, by uniqueness, they are equal, which implies the given *Resolvent Identity* (3.1), and also $R(s,s) = E$, the identity mapping. Hence $R(s,t)$ is bijective, that is, an automorphism of $\mathbb{K}^n$. □

Theorem 3.1.2. *All integrals of* (1.4) *form an n-dimensional linear space over the field $\mathbb{K}$.*

Proof. The linearity of the space of integrals $\mathcal{I}$ is obvious. Let t_0 be a point in I. We show that the mapping

$$\Psi : x \mapsto R(t,t_0)(x)$$

is an isomorphism of $\mathbb{K}^n$ onto $\mathcal{I}$. Linearity is obvious, and, by the identity

$$\frac{d}{dt}R(t,t_0)(x) = \frac{d}{dt}\Phi(t,t_0,x) = A(t)\Phi(t,t_0,x) = A(t)R(t,t_0)(x)\,,$$

the image of Ψ is included in $\mathcal{I}$. If $\Psi(x) = 0$, then we have $R(t,t_0)(x) = 0$ for each t in I. In particular, for $t = t_0$ it follows $x = R(t_0,t_0)(x) = 0$, hence Ψ is an isomorphism. Finally, if φ is an arbitrary integral of (1.4), then, by uniqueness, we have

$$\varphi(t) = R(t,t_0)(\varphi(t_0))$$

for each t in I, hence Ψ is surjective. □

The following theorem can be verified by direct computation.

Theorem 3.1.3. *The complete solution of the Cauchy problem* (1.4), (1.2) *is*

$$\varphi(t) = R(t,t_0)(x_0)\,,$$

and the complete solution of the Cauchy problem (1.3), (1.2) *is*

$$\varphi(t) = R(t,t_0)(x_0) + \int_{t_0}^{t} R(t,s)\big(b(s)\big)ds$$

for each t in I.

An arbitrary basis of the space of all integrals of the homogeneous equation corresponding to (1.3) is called a *fundamental system* of (1.3). The determinant of the matrix U formed by the column vectors $\varphi_1, \varphi_2, \ldots, \varphi_n$, where $\varphi_1, \varphi_2, \ldots, \varphi_n$ are arbitrary integrals of (1.4), is called the *Wronskian* of the integrals $\varphi_1, \varphi_2, \ldots, \varphi_n$:

$$W[\varphi_1, \varphi_2, \ldots, \varphi_n](t) = W(t) = \det U(t)\,.$$

It is easy to see that for all t_0, t in I the identity

$$U(t) = U(t_0)R(t,t_0)$$

holds, as the columns of both matrices on the two sides of this equation are complete solutions of (1.4) with the same initial conditions. Hence, by the Product Rule for determinants, we have

$$\det U(t) = \det U(t_0) \det R(t,t_0)\,.$$

As $\det R(t, t_0) \neq 0$, the Wronskian of n integrals is either identically zero, or never zero, hence n integrals of (1.4) form a fundamental system if and only if their Wronskian is different from zero at some point. For instance, it is obvious that for any fixed t_0 in I the columns of the resolvent $R(t, t_0)$ form a fundamental system of (1.4).

We recall that for an $n \times n$ matrix A the *trace* of A is defined as the sum of the elements in the main diagonal, and it is denoted by $\operatorname{Tr} A$. We have the following remarkable theorem.

Theorem 3.1.4. *(Liouville) Let R be the resolvent of* (1.4). *Then*

$$\det R(t, t_0) = \exp \int_{t_0}^{t} \operatorname{Tr} A(s) ds$$

holds for each t in I.

Proof. By the differentiation rule for determinants we have immediately that the function $t \mapsto \det R(t, t_0)$ is a solution on I of the Cauchy problem

$$x' = \operatorname{Tr} A(t) x$$
$$x(t_0) = 1\,,$$

and the same holds for the function $t \mapsto \exp \int_{t_0}^{t} \operatorname{Tr} A(s) ds$. By uniqueness, the statement follows. □

If we know a fundamental system $\varphi_1, \varphi_2, \ldots, \varphi_n$ of (1.3), then we can determine all integrals of the equation in the following way. As above, we denote by U the matrix having the columns $\varphi_1, \varphi_2, \ldots, \varphi_n$, and we look for an integral of (1.3) in the form

$$\varphi_p(t) = U(t) c(t)\,,$$

where $c : I \to \mathbb{K}^n$ is an unknown differentiable function. As

$$\varphi_p'(t) = U'(t)c(t) + U(t)c'(t) = A(t)U(t)c(t) + U(t)c'(t)\,,$$

hence φ_p is a solution of (1.3) if and only if

$$U(t)c'(t) = b(t)$$

holds for each t in I which implies

$$c(t) = \int_{t_0}^{t} U(s)^{-1} b(s) ds$$

for some t_0 in I. Now it is obvious that for each integral φ of (1.3) the function $\varphi - \varphi_p$ is an integral of (1.4), hence

$$\varphi = \varphi_p + c_1\varphi_1 + c_2\varphi_2 + \cdots + c_n\varphi_n$$

holds with some scalars $c_1, c_2, \ldots, c_n$ in $\mathbb{K}$. This method is called the method of *variation of the constants.* Clearly, this is the generalization of the method with the same name that we introduced in Section 2.5.

3.2 Problems

1. Find the resolvent of the following linear differential equation system:

$$\begin{aligned} x' &= y \\ y' &= 0 \end{aligned}$$

2. Find the resolvent of the following linear differential equation system:

$$\begin{aligned} x' &= \frac{2}{t}x \\ y' &= \frac{3}{t}y \end{aligned}$$

3.3 Linear differential equations with constant coefficients

If $t \mapsto A(t)$ is constant in (1.4), then we call (1.4) a *linear differential equation with constant coefficients.* In this case (1.4) has the following form:

$$x' = Ax\,, \tag{3.2}$$

where A is a linear operator of the linear space $\mathbb{K}^n$. Obviously, every complete solution of (3.2) is defined on the whole $\mathbb{R}$.

Theorem 3.3.1. *The resolvent R of (3.2) satisfies*

$$R(t,s) = \exp A(t-s)$$

for all t, s in $\mathbb{R}$.

Proof. The statement is obvious by the observation that for each fixed real s the columns of the functions $t \mapsto \exp A(t-s)$ and $t \mapsto R(t,s)$ are complete solutions of the same Cauchy problem. □

Theorem 3.3.2. *Let $\mathbb{K} = \mathbb{C}$. Any complex integral φ of the differential equation (3.2) has the form*

$$\varphi(t) = \sum_{k=1}^{q} e^{\lambda_k t} P_k(t)\,,$$

where $\lambda_1, \lambda_2, \ldots, \lambda_q$ are all the different eigenvalues of the matrix A with multiplicities $n_1, n_2, \ldots, n_q$, respectively, further P_k is a polynomial of degree at most $n_k - 1$ with coefficients in $\mathbb{C}^n$.

Proof. The eigenspace X_k corresponding to the eigenvalue λ_k of the matrix A consists of those vectors x of $\mathbb{K}^n$ for which

$$(A - \lambda_k E)^{n_k}(x) = 0$$

holds, further $\mathbb{C}^n$ is the direct sum of the subspaces $X_1, X_2, \ldots, X_q$. If φ is an arbitrary complex integral of (3.2), then we let

$$\varphi(0) = x_1 + x_2 + \cdots + x_q\,,$$

where x_k is an element of X_k for $k = 1, 2, \ldots, q$. Then for each real t we have

$$\varphi(t) = R(t,0)(\varphi(0)) = \exp At(\varphi(0)) = \exp At\Big(\sum_{k=1}^{q} x_k\Big)$$

$$= \sum_{k=1}^{q} \exp At(x_k) = \sum_{k=1}^{q} e^{\lambda_k t} \exp[(A - \lambda_k E)t](x_k)\,.$$

On the other hand, as x_k belongs to X_k, hence

$$\exp[(A - \lambda_k E)t](x_k) = \sum_{j=1}^{n_k - 1} \frac{t^j}{j!}(A - \lambda_k E)^j(x_k) = P_k(t)\,,$$

where P_k is a polynomial of degree at most $n_k - 1$ with coefficients in $\mathbb{C}^n$. The theorem is proved. □

We note that the exponential polynomial φ in the above theorem will not be a solution of (3.2) for an arbitrary polynomial P_k. Practically we can determine the possible polynomials P_k by using the method of "indetermined coefficients", that is, having the eigenvalue λ_k we substitute the function $t \mapsto e^{\lambda_k t} P_k(t)$ with some polynomial of degree at most $n_k - 1$ with unknown coefficients in $\mathbb{C}^n$ into the equation. Then we get a system of linear equations for the coefficients of P_k, from which we can express the coefficients by the free term, where the latter is the solution of the homogeneous system of linear equations

$$(A - \lambda_k E)(x) = 0\,.$$

If A has real coefficients, then we can apply a similar, somewhat more complicated method to find real fundamental system of the equation. The details are left to the reader.

3.4 Problems

1. Solve the following linear sytems:

   i)
$$x' = x + y - 4$$
$$y' = 3y - x$$

   ii)
$$x' = -5y$$
$$y' = 2x + 2y$$

   iii)
$$x' = z - x - y$$
$$y' = x - y - z$$
$$z' = -y$$

2. Solve the linear homogeneous system $y' = Ay$ with the following A:

   i) $A = \begin{pmatrix} -1 & 2 & 3 \\ 0 & 1 & 6 \\ 0 & 0 & -2 \end{pmatrix}$

   ii) $A = \begin{pmatrix} 0 & 2 & 2 \\ 2 & 0 & 2 \\ 2 & 2 & 0 \end{pmatrix}$

   iii) $A = \begin{pmatrix} -1 & 2 & 2 \\ 2 & -1 & 2 \\ 2 & 2 & -1 \end{pmatrix}$

   iv) $A = \begin{pmatrix} 3 & -1 & -1 \\ -2 & 3 & 2 \\ 4 & -1 & -2 \end{pmatrix}$

3. Solve the Cauchy problem $y' = Ax, y(0) = y_0$ with the following A, y_0:

   i) $A = \begin{pmatrix} -7 & 4 \\ -6 & 7 \end{pmatrix}$, $y_0 = \begin{pmatrix} 2 \\ -4 \end{pmatrix}$

   ii) $A = \frac{1}{6}\begin{pmatrix} 7 & 2 \\ -2 & 2 \end{pmatrix}$, $y_0 = \begin{pmatrix} 0 \\ -3 \end{pmatrix}$

   iii) $A = \begin{pmatrix} 21 & -12 \\ 24 & -15 \end{pmatrix}$, $y_0 = \begin{pmatrix} 5 \\ 3 \end{pmatrix}$

iv) $A = \begin{pmatrix} -7 & 4 \\ -6 & 7 \end{pmatrix}$, $y_0 = \begin{pmatrix} -1 \\ 7 \end{pmatrix}$

v) $A = \frac{1}{6}\begin{pmatrix} 1 & 2 & 0 \\ 4 & -1 & 0 \\ 0 & 0 & 3 \end{pmatrix}$, $y_0 = \begin{pmatrix} 4 \\ 7 \\ 1 \end{pmatrix}$

vi) $A = \frac{1}{3}\begin{pmatrix} 2 & -2 & 3 \\ -4 & 4 & 3 \\ 2 & 1 & 0 \end{pmatrix}$, $y_0 = \begin{pmatrix} 1 \\ 1 \\ 5 \end{pmatrix}$

vii) $A = \begin{pmatrix} 6 & -3 & -8 \\ 2 & 1 & -2 \\ 3 & -3 & -5 \end{pmatrix}$, $y_0 = \begin{pmatrix} 0 \\ -1 \\ -1 \end{pmatrix}$

4. Find the general solution of the following systems:

i) $y' = \begin{bmatrix} 3 & 4 \\ -1 & 7 \end{bmatrix} y$

ii) $y' = \begin{bmatrix} 0 & -1 \\ 1 & -2 \end{bmatrix} y$

iii) $y' = \begin{bmatrix} -7 & 4 \\ -1 & -11 \end{bmatrix} y$

iv) $y' = \begin{bmatrix} 3 & 1 \\ -1 & 1 \end{bmatrix} y$

v) $y' = \frac{1}{3}\begin{bmatrix} 1 & 1 & -3 \\ -4 & -4 & 3 \\ -2 & 1 & 0 \end{bmatrix} y$

vi) $y' = \begin{bmatrix} -1 & 1 & -1 \\ -2 & 0 & 2 \\ -1 & 3 & -1 \end{bmatrix} y$

vii) $y' = \begin{bmatrix} 4 & -2 & -2 \\ -2 & 3 & -1 \\ 2 & -1 & 3 \end{bmatrix} y$

viii) $y' = \begin{bmatrix} 6 & -5 & 3 \\ 2 & -1 & 3 \\ 2 & 1 & 1 \end{bmatrix} y$

5. Solve the Cauchy problem $y' = Ay, y(0) = y_0$ with the following A, y_0:

i) $A = \begin{bmatrix} -1 & -4 & -1 \\ 3 & 6 & 1 \\ -3 & -2 & 3 \end{bmatrix}$, $y_0 = \begin{bmatrix} -2 \\ 1 \\ 3 \end{bmatrix}$

ii) $A = \begin{bmatrix} 4 & -8 & -4 \\ -3 & -1 & -3 \\ 1 & -1 & 9 \end{bmatrix}, y_0 = \begin{bmatrix} -4 \\ 1 \\ -3 \end{bmatrix}$

iii) $A = \begin{bmatrix} -5 & -1 & 1 \\ -7 & 1 & 13 \\ -4 & 0 & 8 \end{bmatrix}, y_0 = \begin{bmatrix} 0 \\ 2 \\ 2 \end{bmatrix}$

3.5 Computation of the exponential matrix

The following theorem provides a simple tool for the computation of the resolvent of equation (3.2).

Theorem 3.5.1. *Let A be an $n \times n$ matrix with complex coefficients, whose all different eigenvalues are the complex numbers $\lambda_1, \lambda_2, \ldots, \lambda_q$ with multiplicities $n_1, n_2, \ldots, n_q$, respectively. Then there exist complex polynomials h_{ij} with the following properties: the degree of h_{ij} is at most $n_i - 1$ $(i = 1, 2, \ldots, q; j = 0, 1, \ldots, n_i - 1)$, further if the power series $\sum_k c_k \lambda^k$ has convergence radius R satisfying $\max_i |\lambda_i| < R$, and $|\lambda| < R$, then, with the notation*

$$f(\lambda) = \sum_k c_k \lambda^k$$

we have

$$f(A) = \sum_{i=1}^{q} \sum_{j=0}^{n_i - 1} f^{(j)}(\lambda_i) h_{ij}(A) . \tag{3.3}$$

The polynomials h_{ij} depend only on the matrix A, and as the equality (3.3) holds "for each function" f, hence by different choices of f for which $f(A)$ can be computed in an elementary way (for instance, polynomials, etc.) we can derive several equations for the matrix polynomials $h_{ij}(A)$. We illustrate this method on a simple example.

Let

$$A = \begin{pmatrix} 2 & 1 \\ 3 & 4 \end{pmatrix} .$$

The characteristic polynomial of A is

$$p(\lambda) = \det(A - \lambda I) = (\lambda - 1)(\lambda - 5) ,$$

where I is the identity matrix, hence the eigenvalues are $\lambda_1 = 1$ and $\lambda_2 = 5$, both with multiplicity 1. Therefore in this case the polynomials $h_{1,0}$ and

$h_{2,0}$ have degree zero, that is, they are constants. Let $h_{1,0} = U$ and $h_{2,0} = V$, then, by the above theorem, we have

$$f(A) = f(1)U + f(5)V \tag{3.4}$$

holds for each function f satisfying the conditions of the theorem. If, for example, $f(\lambda) = \lambda - 1$, then by $f(1) = 0$ and $f(5) = 4$ it follows

$$f(A) = A - E = 4V\,,$$

and with the choice $f(\lambda) = \lambda - 5$ we have $f(1) = -4$ and $f(5) = 0$, hence

$$f(A) = A - 5E = -4U\,,$$

which implies

$$U = \frac{1}{4}\begin{pmatrix} 3 & -1 \\ -3 & 1 \end{pmatrix} \qquad \text{and} \qquad V = \frac{1}{4}\begin{pmatrix} 1 & 1 \\ 3 & 3 \end{pmatrix}.$$

This means that equality (3.4) holds for each complex valued function f which can be expanded into a power series, convergent on a circle centered at the origin and having convergence radius greater than 5. For such a function f we have

$$f(A) = \frac{f(1)}{4}\begin{pmatrix} 3 & -1 \\ -3 & 1 \end{pmatrix} + \frac{f(5)}{4}\begin{pmatrix} 1 & 1 \\ 3 & 3 \end{pmatrix}.$$

In particular, with the choice $f(\lambda) = \exp\lambda$ we have $f(1) = e$ and $f(5) = e^5$, hence

$$e^A = \frac{e}{4}\begin{pmatrix} 3 & -1 \\ -3 & 1 \end{pmatrix} + \frac{e^5}{4}\begin{pmatrix} 1 & 1 \\ 3 & 3 \end{pmatrix}.$$

Using our previous results we can find all integrals of the first order system of homogeneous linear differential equations

$$\begin{aligned} x' &= 2x + y \\ y' &= 3x + 4y\,. \end{aligned}$$

The matrix of this system is the matrix A we have studied above, hence the resolvent is

$$R(t, t_0) = \exp A(t - t_0)$$

for each real t, t_0. By the above considerations the matrix on the right hand side can be computed with the choice

$$f(\lambda) = \exp\lambda(t - t_0)\,,$$

and we have

$$\exp \lambda(t - t_0) = \frac{e^{t-t_0}}{4} \begin{pmatrix} 3 & -1 \\ -3 & 1 \end{pmatrix} + \frac{e^{5(t-t_0)}}{4} \begin{pmatrix} 1 & 1 \\ 3 & 3 \end{pmatrix}.$$

The columns of the resolvent form a fundamental system of our system of differential equations and they have the form

$$x_1(t) = \frac{3}{4}e^{t-t_0} + \frac{1}{4}e^{5(t-t_0)}, \qquad y_1(t) = -\frac{3}{4}e^{t-t_0} + \frac{3}{4}e^{5(t-t_0)},$$

$$x_2(t) = -\frac{1}{4}e^{t-t_0} + \frac{1}{4}e^{5(t-t_0)}, \qquad y_2(t) = \frac{1}{4}e^{t-t_0} + \frac{3}{4}e^{5(t-t_0)}.$$

This can be reformulated in the following way: the general solution of our system of homogeneous linear differential equations is

$$\begin{aligned} x(t) &= ae^t + be^{5t}, \\ y(t) &= -ae^t + 3be^{5t}, \end{aligned}$$

where a, b are arbitrary real or complex numbers, depending on if we are looking for real or complex solutions. The complete solution of the system satisfying the initial conditions $x(t_0) = x_0$, $y(t_0) = y_0$ is

$$\begin{aligned} x(t) &= x_0 \cdot x_1(t) + y_0 \cdot x_2(t), \\ y(t) &= x_0 \cdot y_1(t) + y_0 \cdot y_2(t). \end{aligned}$$

Now we consider the first order inhomogeneous linear system of differential equations

$$\begin{aligned} x' &= 2x + y + 1 \\ y' &= 3x + 4y - t. \end{aligned}$$

The corresponding homogeneous system is the one we have solved above. A particular solution of the inhomogeneous system can be determined using the method of variation of the constants. From the general solution of the homogeneous system we let

$$\begin{aligned} x_p(t) &= a(t)e^t + b(t)e^{5t}, \\ y_p(t) &= -a(t)e^t + 3b(t)e^{5t}, \end{aligned}$$

where a, b are unknown differentiable functions. Substitution into the inhomogeneous system shows that x_p, y_p is a solution of the inhomogeneous system if and only if the equations

$$\begin{aligned} 1 &= a'(t)e^t + b'(t)e^{5t}, \\ -t &= -a'(t)e^t + 3b'(t)e^{5t} \end{aligned}$$

hold for each real t. This is a homogeneous linear system of equations with the unique solution

$$a'(t) = \frac{1}{4}(3+t)e^{-t},$$
$$b'(t) = \frac{1}{4}(1-t)e^{-5t},$$

from which one derives by integration

$$a(t) = -e^{-t} - \frac{1}{4}te^{-t},$$
$$b(t) = -\frac{1}{25}e^{-5t} + \frac{1}{20}te^{-5t},$$

where we have chosen the integration constants to be zero. By substitution, we get a particular solution of the inhomogeneous system in the form

$$x_p(t) = -\frac{26}{25} - \frac{1}{5}t,$$
$$y_p(t) = \frac{22}{25} + \frac{2}{5}t.$$

Finally, the general solution of the inhomogeneous system can be obtained as the sum of the general solution of the homogeneous system and the particular solution:

$$x(t) = ae^{t} + be^{5t} - \frac{26}{25} - \frac{1}{5}t,$$
$$y(t) = -ae^{t} + 3be^{5t} + \frac{22}{25} + \frac{2}{5}t,$$

where a, b are arbitrary real or complex constants, depending on whether we are looking for real or complex solutions. The complete solution of the inhomogeneous system satisfying the initial conditions $x(t_0) = x_0$, $y(t_0) = y_0$ can be obtained by an appropriate choice of the constants a, b.

3.6 Problems

1. Find e^{tA} for the following matrices:

   i)

$$A = \begin{pmatrix} -2 & -4 \\ 1 & 2 \end{pmatrix}$$

   ii)

$$A = \begin{pmatrix} -1 & -5 \\ -1 & 3 \end{pmatrix}$$

iii)

$$A = \begin{pmatrix} 0 & 0 & 1 \\ 0 & 0 & 0 \\ 1 & 0 & 0 \end{pmatrix}$$

iv)

$$A = \begin{pmatrix} 2 & 0 & -1 \\ 0 & 2 & 0 \\ 0 & 0 & 2 \end{pmatrix}$$

2. Find $e^{\frac{\pi}{3}A}b$ for

$$A = \begin{pmatrix} 1 & -2 \\ 1 & -1 \end{pmatrix}$$

and

$$b = \begin{pmatrix} 2 \\ 4 \end{pmatrix}.$$

3. Find e^{tA} for the following matrices:

i)

$$A = \begin{pmatrix} 2 & -1 \\ 4 & -2 \end{pmatrix}$$

ii)

$$A = \begin{pmatrix} 3 & 4 \\ -1 & -1 \end{pmatrix}$$

iii)

$$A = \begin{pmatrix} 1 & -2 \\ 1 & -1 \end{pmatrix}$$

4. Find the general solution of the following systems of differential equations:

i)

$$\begin{aligned} y_1' &= 2y_1 + 4y_2 \\ y_2' &= 4y_1 + 2y_2 \end{aligned}$$

ii)

$$\begin{aligned} y_1' &= 3y_1 - y_2 - y_3 \\ y_2' &= -2y_1 + 3y_2 + 2y_3 \\ y_3' &= 4y_1 - y_2 - 2y_3 \end{aligned}$$

iii)

$$\begin{aligned} y_1' &= -3y_1 + 2y_2 + 2y_3 \\ y_2' &= 2y_1 - 3y_2 + 2y_3 \\ y_3' &= 2y_1 + 2y_2 - 3y_3 \end{aligned}$$

iv)

$$\begin{aligned} y_1' &= -5y_1 + 5y_2 + 4y_3 \\ y_2' &= -8y_1 + 7y_2 + 6y_3 \\ y_3' &= y_1 \end{aligned}$$

Chapter 4

FUNCTIONAL DEPENDENCE, INDEPENDENCE

4.1 Functional independence

Throughout this chapter n will denote a positive integer. Let $\Omega \subseteq \mathbb{R}^n$ be a non-empty open set, k a positive integer, and x_0 a point in Ω, further let $\varphi_j : \Omega \to \mathbb{R}$ be continuously differentiable functions ($j = 1, 2, \ldots, k$). We say that the functions φ_j are *functionally independent* at the point x_0, if the rank of the matrix $\big(\partial_j \varphi_i\big)(x_0)$ ($i = 1, 2, \ldots, k$; $j = 1, 2, \ldots, n$) is equal to k, the number of these functions. Observe that this matrix is exactly the derivative at x_0 of the function $\varphi : \Omega \to \mathbb{R}^k$ defined by $\varphi = (\varphi_1, \varphi_2, \ldots, \varphi_k)$, the so-called *Jacobian* which can also be denoted by $\frac{\partial(\varphi_1, \varphi_2, \ldots, \varphi_k)}{\partial(x_1, x_2, \ldots, x_n)}$. Otherwise we say that the functions φ_j are *functionally dependent* at the point x_0. We can simply say *independent* or *dependent*, if this does not lead to confusion. If these functions are independent, or dependent at each point of Ω, then we say that they are *independent*, or *dependent on* Ω, respectively.

Theorem 4.1.1. *Let $\Omega \subseteq \mathbb{R}^n$ be an non-empty open set, $k < n$ a positive integer, and let the continuously differentiable functions $\varphi_1, \varphi_2, \ldots, \varphi_k : \Omega \to \mathbb{R}$ be independent at the point x_0 in Ω. Then there exists a neighborhood U of x_0 in Ω, and there are continuously differentiable functions $\varphi_j : U \to \mathbb{R}$ for $j = k+1, k+2, \ldots, n$ such that $\varphi_1, \varphi_2, \ldots, \varphi_n$ are independent on U.*

Proof. As the rows of the $k \times n$ matrix $\big(\partial_j \varphi_i\big)(x_0)$ with $j = 1, 2, \ldots, n$ and $i = 1, 2, \ldots, k$ are linearly independent, hence, by adding appropriate rows, this matrix can be enlarged to a regular $n \times n$ matrix. The new rows can be considered as partial derivatives of linear functions $\varphi_{k+1}, \varphi_{k+2}, \ldots, \varphi_n$. As the determinant of the quadratic matrix $\big(\partial_j \varphi_i\big)(x_0)$ with $i = 1, 2, \ldots, n$ and $j = 1, 2, \ldots, n$ is different from zero, hence x_0 has a neighborhood U in

Ω such that for each point x in U the determinant of the matrix $(\partial_j \varphi_i)(x)$ with $i = 1, 2, \ldots, n$ and $j = 1, 2, \ldots, n$ is different from zero, hence the functions $\varphi_1, \varphi_2, \ldots, \varphi_n$ are independent on U. □

We remark that, by definition, the continuously differentiable functions $\varphi_1, \varphi_2, \ldots, \varphi_n : \Omega \to \mathbb{R}$ are dependent on the open set $\Omega \subseteq \mathbb{R}^n$ if and only if $\det \varphi'(x) = 0$ holds at each point x in Ω, where $\varphi = (\varphi_1, \varphi_2, \ldots, \varphi_n)$.

Theorem 4.1.2. *Let $\Omega \subseteq \mathbb{R}^n$ be an arbitrary non-empty open set, further let $\varphi_j : \Omega \to \mathbb{R}$ be continuously differentiable functions for $j = 1, 2, \ldots, n$, and we define $\varphi = (\varphi_1, \varphi_2, \ldots, \varphi_n)$. The functions φ_j are dependent on Ω if and only if for each closed ball B in Ω the set $\varphi(B)$ is nowhere dense.*

Proof. Suppose that for some x in Ω we have $\det \varphi'(x) \neq 0$, that is, the functions $\varphi_1, \varphi_2, \ldots, \varphi_n$ are not functionally dependent on Ω. Then, by the Inverse Function Theorem, there exists a neighborhood of x whose image at φ is a neighborhood of $\varphi(x)$, hence it is impossible that the image of all closed balls in Ω is nowhere dense.

Conversely, assume that $\det \varphi'(x) = 0$ holds for each x in Ω. It is well-known from the theory of multiple Riemann integral that if B is a closed ball in Ω, then $\varphi(B)$ is Jordan measurable and its Jordan measure is zero. This implies that $\varphi(B)$ is nowhere dense, and the theorem is proved. □

Theorem 4.1.3. *Let $\Omega \subseteq \mathbb{R}^n$ be a non-empty open set, and suppose that the functions $\varphi_j : \Omega \to \mathbb{R}$ are continuously differentiable for $j = 1, 2, \ldots, n$. The functions φ_j are functionally dependent on Ω if and only if for each closed ball B in Ω there is a continuously differentiable function $F : \mathbb{R}^n \to \mathbb{R}$ which does not vanish identically on any non-empty open set, and it satisfies*

$$F\big(\varphi_1(x), \varphi_2(x), \ldots, \varphi_n(x)\big) = 0$$

for each x in B.

Proof. Let $\varphi = (\varphi_1, \varphi_2, \ldots, \varphi_n)$. First we suppose that for some closed ball B in Ω the set $\varphi(B)$ is nowhere dense. Then each continuous function $F : \mathbb{R}^n \to \mathbb{R}$ which vanishes on $\varphi(B)$ also vanishes on the closure of $\varphi(B)$, and this latter has an interior point, hence F vanishes on some non-empty open set.

For the converse we prove the following, somewhat stronger statement: if S is a nowhere dense subset in $\mathbb{R}^m$, then there exists a continuously differentiable function which does not vanish identically on any non-empty open set, but it vanishes at every point of S. By Theorem 4.1.2, this will imply our statement.

We consider in $\mathbb{R}^m$ the family T_0 of all closed rectangular boxes, whose sides are of unit length and whose vertices have integer coordinates. We define the function f_0 on $\mathbb{R}^m$ in the following manner:

$$f_0(x) = \frac{1}{2}\sin^2 \pi x_1 \cdot \sin^2 \pi x_2 \cdot \dots \cdot \sin^2 \pi x_m\,,$$

whenever $x = (x_1, x_2, \dots, x_m)$ belongs to some element in T_0 which does not intersect S, otherwise we let $f_0(x) = 0$. It is easy to see that $f_0 : \mathbb{R}^m \to \mathbb{R}$ is continuous, and it is continuously partially differentiable with respect of each variable. Now we cut in half the sides of each box in T_0, and we denote by T_1 the system of closed rectangular boxes obtained in this way. We define the function f_1 in the following manner

$$f_1(x) = \frac{1}{2^{2m+1}}\sin^2 2\pi x_1 \cdot \sin^2 2\pi x_2 \cdot \dots \cdot \sin^2 2\pi x_m\,,$$

whenever $x = (x_1, x_2, \dots, x_m)$ belongs to some element in T_0 which does not intersect S, otherwise we let $f_1(x) = 0$. Continuing this process we obtain the sequence $f_0, f_1, \dots$ of continuous and continuously partially differentiable functions which satisfy the inequalities

$$0 \leqslant f_k(x) \leqslant \frac{1}{2^{2mk+1}}$$

and

$$|\partial_i f_k(x)| \leqslant \frac{\pi}{2^{(2m-1)k}}$$

for $k = 0, 1, \dots,$ $i = 1, 2, \dots, m$ and for each x in $\mathbb{R}^m$. Consequently, the function series $\sum_k f_k$ is uniformly convergent on $\mathbb{R}^m$. Let F denote its sum, then F is continuous, and $F(x) = 0$ for each x in S. On the other hand, if Ω_0 is a non-empty open set in $\mathbb{R}^m$, then there exists a non-empty open set $\Omega_1 \subseteq \Omega_0$ such that $\Omega_1 \cap S = \varnothing$, further there exists a positive integer N such that some box of T_N is a subset of Ω_1. At each interior point x of this box we have $f_N(x) > 0$, hence $F(x) > 0$. It follows that F does not vanish identically on any non-empty open set. Finally, for $i = 1, 2, \dots, m$ the series $\sum_k \partial_i f_k$ converges uniformly on $\mathbb{R}^m$, too, hence F is continuously partially differentiable with respect to each variable. We conclude that F is continuously differentiable and the theorem is proved. □

4.2 Functional expressibility

Let $\Omega \subseteq \mathbb{R}^n$ be a non-empty open set, k a positive integer, further let $\psi, \varphi_j : \Omega \to \mathbb{R}$ $(j = 1, 2, \ldots, k)$ be continuously differentiable functions. We say that ψ is *functionally expressible*, or simply *expressible* on Ω with the functions φ_j, if each point in Ω has a neighborhood $U \subseteq \Omega$, further there exists an open set $E \subseteq \mathbb{R}^k$ and a continuously differentiable function $\Phi : E \to \mathbb{R}$ such that for each x in U the point $\big(\varphi_1(x), \varphi_2(x), \ldots, \varphi_k(x)\big)$ is in E, and we have

$$\psi(x) = \Phi\big(\varphi_1(x), \varphi_2(x), \ldots, \varphi_k(x)\big) .$$

Let x_0 be arbitrary in Ω, and suppose that in some neighborhood of x_0 the function ψ is expressible with the functions φ_j. Then, obviously, $\psi, \varphi_1, \varphi_2, \ldots, \varphi_k$ are functionally dependent at x_0. The following theorem is a kind of converse of this statement.

Theorem 4.2.1. *Let $\Omega \subseteq \mathbb{R}^n$ be a non-empty open set, k a positive integer, and let $\psi, \varphi_j : \Omega \to \mathbb{R}$ $(j = 1, 2, \ldots, k)$ be continuously differentiable functions. If the functions φ_j are functionally independent on Ω, and the functions $\psi, \varphi_1, \varphi_2, \ldots, \varphi_k$ are functionally dependent on Ω, then ψ is functionally expressible on Ω with the functions φ_j $(j = 1, 2, \ldots, k)$.*

Proof. Let x_0 be a point in Ω, and let $U \subseteq \Omega$ be a neighborhood of x_0 for which there are continuously differentiable functions $\varphi_{k+1}, \varphi_{k+2}, \ldots, \varphi_n$ on U such that $\varphi_1, \varphi_2, \ldots, \varphi_n$ are functionally independent on U. Let $\varphi = (\varphi_1, \varphi_2, \ldots, \varphi_n)$ and $z_0 = \varphi(x_0)$. As $\det \varphi'(x_0) \neq 0$, hence, by the Inverse Function Theorem, there exists a neighborhood $V \subseteq U$ of x_0 and a connected neighborhood W of z_0 such that φ maps V bijectively onto W, and the inverse is continuously differentiable. Let φ^{-1} denote the inverse of the restriction of φ onto V, and for each z in W we let

$$\Psi(z) = \psi\big(\varphi^{-1}(z)\big) .$$

Then Ψ is a continuously differentiable function. Let E denote the set of all points $(z_1, z_2, \ldots, z_k)$ in $\mathbb{R}^k$ such that there exists a point $(z_{k+1}, z_{k+2}, \ldots, z_n)$ in $\mathbb{R}^{n-k}$ for which $(z_1, z_2, \ldots, z_n)$ belongs to W. Obviously, E is an open set which contains the point $\big(\varphi_1(x_0), \varphi_2(x_0), \ldots, \varphi_k(x_0)\big)$. We show that if $(z_1, \ldots, z_k, z_{k+1}, \ldots, z_n)$ and $(z_1, \ldots, z_k, \overline{z}_{k+1}, \ldots, \overline{z}_n)$ belongs to W, then we have

$$\Psi(z_1, \ldots, z_k, z_{k+1}, \ldots, z_n) = \Psi(z_1, \ldots, z_k, \overline{z}_{k+1}, \ldots, \overline{z}_n) ,$$

in other words, Ψ depends only on its first k variables. To prove this it is enough to show, by the connectedness of W, that $\partial_j \Psi(z) = 0$, whenever z is in W for $j = k+1, k+2, \ldots, n$.

It is clear that we have for each z in W

$$\partial_j \Psi(z) = \sum_{i=1}^{n} \partial_i \psi\big(\varphi^{-1}(z)\big) \cdot \partial_j \big(\varphi_i^{-1}(z)\big)\,,$$

where φ_i^{-1} denotes the i-th component of the function φ^{-1} $(i = 1, 2, \ldots, n)$. On the other hand, for each x in V we have

$$\partial_i \psi(x) = \sum_{s=1}^{k} a_s(x) \cdot \partial_i \varphi_s(x)$$

for $i = 1, 2, \ldots, n$, as the functions $\varphi_1, \varphi_2, \ldots, \varphi_k$ are independent and $\psi, \varphi_1, \varphi_2, \ldots, \varphi_k$ are dependent. Here $a_s : V \to \mathbb{R}$ are some unknown functions for $s = 1, 2, \ldots, k$. We conclude for $j = k+1, k+2, \ldots, n$

$$\partial_j \Psi(z) = \sum_{s=1}^{k} a_s\big(\varphi^{-1}(z)\big) \cdot \sum_{i=1}^{n} \partial \varphi_s\big(\varphi^{-1}(z)\big) \cdot \partial_j (\varphi^{-1})_i(z)$$

$$= \sum_{s=1}^{k} a_s\big(\varphi^{-1}(z)\big) \cdot \partial_j (\varphi \circ \varphi^{-1})_s(z) = 0\,,$$

as $j > k$. This proves our statement.

Finally, for each $(z_1, z_2, \ldots, z_k)$ in E we have

$$\Phi(z_1, z_2, \ldots, z_k) = \Psi(z_1, z_2, \ldots, z_k, z_{k+1}, \ldots, z_n)\,,$$

where $(z_1, z_2, \ldots, z_k, z_{k+1}, \ldots, z_n)$ belongs to W. Obviously, $\Phi : E \to \mathbb{R}$ is continuously differentiable, the point $\big(\varphi_1(x), \varphi_2(x), \ldots, \varphi_k(x)\big)$ belongs to E for each x in V, and

$$\psi(x) = \psi\big(\varphi^{-1}(z)\big) = \Psi(z) = \Psi\big(\varphi_1(x), \varphi_2(x), \ldots, \varphi_n(x)\big)$$

$$= \Phi\big(\varphi_1(x), \varphi_2(x), \ldots, \varphi_k(x)\big)\,,$$

which was to be proved. $\square$

4.3 First integrals

Let $\Omega \subseteq \mathbb{R} \times \mathbb{R}^n$ be a non-empty open set, and $f : \Omega \to \mathbb{R}^n$ a continuous function. We say that ψ is a *first integral* of the differential equation

$$x' = f(t, x) \tag{4.1}$$

on a non-empty open subset $\Omega_0 \subseteq \Omega$, if $\psi : \Omega_0 \to \mathbb{R}$ is a continuously differentiable function, and whenever $x : I \to \mathbb{R}^n$ is a solution of (1.1) such that the point $\big(t, x(t)\big)$ belongs to Ω_0 for each t in I, the function $t \mapsto \varphi\big(t, x(t)\big)$ is constant on I. In other words, ψ is a first integral of (1.1) on Ω_0, if it is constant along the graph of each solution, if that graph lies in Ω_0. The first integral ψ is called *autonomous first integral*, if $\partial_1 \psi$ is identically zero, that is, ψ does not depend on the variable t.

It is obvious that if the functions $\psi_1, \psi_2, \ldots, \psi_k$ are first integrals of (1.1) on Ω_0, and ψ is expressible on Ω_0 with the functions $\psi_1, \psi_2, \ldots, \psi_k$, then ψ is a first integral of (1.1) on Ω_0, too. It is easy to see that any finite set of first integrals of (1.1) are functionally dependent, if their number is greater than n. The basic question is if each point of Ω has a neighborhood in which there exist n functionally independent first integrals of (1.1). In the subsequent paragraphs we show that this is the case under quite general natural conditions on (1.1).

Let $\Omega \subseteq \mathbb{R} \times \mathbb{R}^n$ be a non-empty open set and let $f : \Omega \to \mathbb{R}^n$ be a continuously differentiable function. Let Φ denote the characteristic function of the differential equation (1.1). We recall that if for each (τ, ξ) in Ω we denote by $I_{(\tau,\xi)}$ the domain of the complete solution $\varphi_{(\tau,\xi)}$ of (1.1) satisfying the initial condition $x(\tau) = \xi$, then the characteristic function φ which is defined for each (τ, ξ) in Ω and t in $I_{(\tau,\xi)}$ by the equation

$$\Phi(t, \tau, \xi) = \varphi_{(\tau,\xi)}(t)\,,$$

is continuously differentiable by the conditions on f. Further, if V denotes the domain of Φ, then for each point (τ, ξ) in V we have

$$\det \partial_3\, \varphi(t, \tau, \xi) \neq 0\,.$$

The following theorem shows that using the characteristic function of the differential equation (1.1) it is possible to construct n functionally independent first integrals of (1.1) in some neighborhood of every point of Ω.

Theorem 4.3.1. *Let $\Omega \subseteq \mathbb{R} \times \mathbb{R}^n$ be a non-empty open set, and $f : \Omega \to \mathbb{R}^n$ a continuously differentiable function. Let Φ denote the characteristic function of the differential equation* (1.1). *Then each point (t_0, x_0) of Ω has a* *neighborhood $U \subseteq \Omega$ such that the components $\psi_1, \psi_2, \dots, \psi_n$ of the restriction Ψ to U of the function $(t,x) \mapsto \Phi(t_0, t, x)$ are functionally independent first integrals on U of* (1.1).

Proof. Let (t_0, x_0) be arbitrary in Ω and we denote by V the domain of the characteristic function Φ. By the previous results, we know that V is an open set. As (t_0, t_0, x_0) belongs to V, hence the point (t_0, x_0) has a neighborhood $U \subseteq \Omega$ such that for each (t,x) in U the point (t_0, t, x) belongs to V. As for each (t,x) in U we have

$$\det \partial_3 \Phi(t_0, t, x) \neq 0\,,$$

hence the rank of the matrix

$$\big(\partial_i \psi_j\big)(t,x) = \big(\partial_2 \Phi(t_0, t, x), \partial_3 \Phi(t_0, t, x)\big)$$

is n, consequently, the functions $\psi_1, \psi_2, \dots, \psi_n$ are independent on U. We show that these functions are first integrals of (1.1) on U.

Let $x : I \to \mathbb{R}^n$ be a solution of (1.1) such that for each t in I the point $\big(t, x(t)\big)$ is in U. We denote by $\overline{x}$ the continuation of x up to the boundary. We show that $\overline{x}$ is defined at t_0. If $\overline{x}$ is defined at some point t_1, then $\big(t_1, \overline{x}(t_1)\big)$ belongs to U, hence $\big(t_0, t_1, \overline{x}(t_1)\big)$ is in V, that is, the complete solution passing through the point $\big(t_1, \overline{x}(t_1)\big)$ is defined at t_0. But this complete solution is $\overline{x}$. Finally, if t is arbitrary in I, then we have

$$\Psi\big(t, x(t)\big) = \Psi\big(t, \overline{x}(t)\big) = \Phi\big(t_0, t, \overline{x}(t)\big) = \overline{x}(t_0)\,,$$

which is constant. The theorem is proved. □

As an example we consider the differential equation

$$x' = x$$
$$y' = y\,.$$

The complete solution satisfying the initial condition $x(\tau) = \xi$, $y(\tau) = \eta$ is given by

$$x(t) = \xi e^{t-\tau}, \qquad y(t) = \eta e^{t-\tau}\,,$$

hence the characteristic function is

$$\Phi(t, \tau, \xi, \eta) = (\xi e^{t-\tau}, \eta e^{t-\tau})\,.$$

Consequently, two functionally independent first integrals of the equation can be given as

$$\psi_1(t, x, y) = x e^{t_0 - t}, \qquad \psi_2(t, x, y) = y e^{t_0 - t}$$

for each t, t_0 in $\mathbb{R}$.

It is easy to see that having n independent first integrals of (1.1) it is possible to describe the complete solution of (1.1) satisfying any initial condition.

4.4 Problems

1. Find a non-constant first integral of the following system of differential equations:
$$\frac{dx}{z} = \frac{dy}{xz} = \frac{dz}{y}.$$
2. Find two functionally independent first integrals of the following system of differential equations:
$$\frac{dx}{z^2 - y^2} = \frac{dy}{z} = -\frac{dz}{y}.$$
3. Find two functionally independent autonomous first integrals of the following system of differential equations:
$$\frac{dx}{y+z} = \frac{dy}{x+z} = \frac{dz}{x+y}.$$
4. Find two functionally independent autonomous first integrals of the following system of differential equations:
$$\frac{dx}{x(y^2+z^2)} = -\frac{dy}{y(x^2+z^2)} = \frac{dz}{z(x^2+y^2)}.$$
5. Find two functionally independent autonomous first integrals of the following system of differential equations:
$$\frac{dx}{x(x+y)} = -\frac{dy}{y(x+y)} = \frac{dz}{(x-y)(2x+2y+z)}.$$
6. Find three functionally independent first integrals of the following system of differential equations:
$$\frac{dx}{dt} = \frac{x-y}{z-t}, \quad \frac{dy}{dt} = \frac{x-y}{z-t}, \quad \frac{dz}{dt} = x - y + 1.$$
7. Show that if $n > 1$ and the right hand side of (1.1) is continuously differentiable, then every point of Ω has a neighborhood on which there are $n-1$ functionally independent autonomous first integrals of (1.1).

Chapter 5

HIGHER ORDER DIFFERENTIAL EQUATIONS

5.1 A reduction principle

Let D be a non-empty open subset in $\mathbb{R} \times \mathbb{K} \times \mathbb{K} \times \cdots \times \mathbb{K}$, where there are n copies of $\mathbb{K}$, and let $f : D \to \mathbb{K}$ be a function. The equation

$$x^{(n)} = f(t, x, x', \ldots, x^{(n-1)}) \tag{5.1}$$

is called an *ordinary explicit differential equation of order* n on the set D, or simply a *differential equation of order* n. The set D, and the function f are called the *domain*, and the *right side* of (5.1), respectively.

We say that the function $\varphi : I \to \mathbb{K}$ is a *solution of* (5.1) *on* I, if I is an open interval in $\mathbb{R}$, φ is an n-times differentiable function, the point

$$(t, \varphi(t), \varphi'(t), \ldots, \varphi^{(n-1)}(t))$$

belongs to D for each t in I and

$$\varphi^{(n)}(t) = f(t, \varphi(t), \varphi'(t), \ldots, \varphi^{(n-1)}(t))$$

holds.

If $(t_0, x_0, x_1, \ldots, x_{n-1})$ is any point of D, then the system of equations

$$x^{(i)}(t_0) = x_i, \qquad (i = 0, 1, \ldots, n-1) \tag{5.2}$$

is called an *initial condition* for (5.2), and the system (5.1), (5.2) is called a *Cauchy problem* or *initial value problem*. We say that φ is a *solution of the Cauchy problem* (5.1), (5.2) *on* I, if φ is a solution of (5.1) on I, I includes t_0 and $\varphi^{(i)}(t_0) = x_i$ holds for $i = 0, 1, \ldots, n-1$.

We say that the Cauchy problem (5.1), (5.2) has a *locally unique* solution, if any two solutions coincide on the intersection of their domains.

A solution of the n-th order differential equation (5.1) or the Cauchy problem (5.1), (5.2) is called a *complete solution*, if it has no proper extension to a solution.

If in (5.1) $D = I \times \mathbb{K} \times \mathbb{K} \times \cdots \times \mathbb{K}$, where I is an open interval in $\mathbb{R}$, further $f : D \to \mathbb{K}$ is continuous and is linear in its second, third, etc., $n+1$-th variable, then (5.1) is called a *linear differential equation of n-th order* on I. Hence the general form of the linear differential equation of n-th order is

$$x^{(n)} + a_1(t)x^{(n-1)} + \cdots + a_n(t)x = a_0(t), \tag{5.3}$$

where $a_i : I \to \mathbb{K}$ is a continuous function ($i = 0, 1, \ldots, n$).

Let $\varphi : I \to \mathbb{K}$ be an n-times differentiable function and let for each t in I

$$\widetilde{\varphi}(t) = (\varphi(t), \varphi'(t), \ldots, \varphi^{(n-1)}(t)) \, .$$

Further, using the notation $y = (y_0, y_1, \ldots, y_{n-1})$, we introduce

$$\widetilde{f}(t, y) = (y_1, y_2, \ldots, y_{n-1}, f(t, y_0, y_1, \ldots, y_{n-1})$$

for each $(t, y_0, y_1, \ldots, y_{n-1}$ in D. It is easy to see that φ is a solution of (5.1) on I if and only if $\widetilde{\varphi}$ is a solution of the first order differential equation

$$x' = \widetilde{f}(t, x)$$

on I. Moreover, using the notation $z_0 = (x_0, x_1, \ldots, x_{n-1})$, the function φ is a solution of the Cauchy problem (5.1), (5.2) on I, if and only if $\widetilde{\varphi}$ is a solution of the Cauchy problem

$$x' = \widetilde{f}(t, x)$$
$$x(t_0) = z_0$$

on I. This observation, called *Transition Principle*, makes it possible to transfer the results of the previous sections to n-th order differential equations. The following theorems are simple consequences of the Transition Principle.

Theorem 5.1.1. *If f is continuous in* (5.1), *then the Cauchy problem* (5.1), (5.2) *has a solution.*

Theorem 5.1.2. *If $f, \partial_2 f, \ldots, \partial_{n+1} f$ are continuous in* (5.1), *then the Cauchy problem* (5.1), (5.2) *has a locally unique solution and a unique complete solution.*

Theorem 5.1.3. *If* (5.1) *is linear on I, then the Cauchy problem* (5.1), (5.2) *has a unique solution on I.*

5.2 Problems

1. Prove Theorem 5.1.1.
2. Prove Theorem 5.1.2.
3. Prove Theorem 5.1.3.

5.3 Intermediate integrals

Besides explicit differential equations of the form (5.1) we may consider the more general implicit differential equation

$$F(t, x, x', \ldots, x^{(n)}) = 0\,, \tag{5.4}$$

where we suppose that F is defined on some non-empty open subset of $\mathbb{R}^{n+2}$ and it has reasonable differentiability properties. The concept of the solution of this implicit equation can be defined easily, the details are left to the reader. In this section we present a method for decreasing the degree of (5.4), that is, to find a similar equation with smaller degree such that every solution of the new equation is a solution of (5.4).

Suppose that we are given the n-th order explicit differential equation (5.4), and assume that we have a relation of the form

$$\Phi(t, x, C_1, C_2, \ldots, C_n) = 0\,, \tag{5.5}$$

where Φ is a given function which is sufficiently many times differentiable, $C_1, C_2, \ldots, C_n$ are real parameters, and (5.5) holds on some open set of the variable (t, x). By differentiating (5.5) n times with respect to t and considering x as a function of t we obtain a system of equations including (5.5). If, by eliminating the parameters from the system obtained, we get a differential equation equivalent to (5.4), then we call (5.5) a *general integral* of (5.4). Supposing that Φ has some additional properties which guarantee that x can be expressed form (5.5), then we have a family of solutions of (5.4) depending on the n parameters $C_1, C_2, \ldots, C_n$. Under reasonable conditions on F in (5.4) this family will include all solutions of (5.4). Although we will not go into the detailed study of such conditions, however, we present some practical methods to find as many solutions as possible.

More generally, suppose that we have a relation of the form

$$\Phi(t, x, x', x'', \ldots, x^{(k)}, C_{k+1}, \ldots, C_n) = 0\,, \tag{5.6}$$

with some $1 \leqslant k \leqslant n-1$. By differentiating (5.5) $n-k$ times with respect to t and considering x as a function of t we obtain a system of equations including (5.5). If, by eliminating the parameters from the system obtained, we get a differential equation equivalent to (5.4), then we call (5.6) an *intermediate integral* of (5.4).

By differentiating (5.6) $n-k$ times with respect to t we have the system of equations

$$\begin{aligned} \partial_1\Phi + \partial_2\Phi \cdot x' + \cdots + \partial_{k+2}\Phi \cdot x^{(k+1)} &= 0 \\ \partial_1^2\Phi + \cdots + \partial_{k+2}\Phi \cdot x^{(k+2)} &= 0 \\ &\cdots \\ &\cdots \\ &\cdots \\ \partial_1^{n-k}\Phi + \cdots + \partial_{k+2}\Phi \cdot x^{(n)} &= 0 \,. \end{aligned}$$

Eliminating the constants $C_{k+1}, \ldots, C_n$ from this system, by assumption, we get the differential equation (5.4). This means that every solution of (5.6) is a solution of (5.4). However, the order of (5.6) is $k < n$, hence we reduced the problem of finding solutions of (5.4) to the problem of finding solutions of a differential equation of smaller order.

We consider the differential equation

$$x^{(n)} = f(t)\,. \tag{5.7}$$

By successive integration we get intermediate integrals and a general integral for this equation in the form

$$\begin{aligned} x^{(n-1)} &= \int_{t_0}^{t} f(s_1)ds_1 + C_1 \\ x^{(n-2)} &= \int_{t_0}^{t}\int_{t_0}^{s_2} f(s_1)ds_1 ds_2 + C_1(t-t_0) + C_2 \\ &\cdots \\ &\cdots \\ &\cdots \\ x &= \int_{t_0}^{t}\int_{t_0}^{s_1}\cdots\int_{t_0}^{s_{n-1}} f(s_n)ds_n ds_{n-1}\ldots ds_1 + \sum_{k=1}^{n}\frac{C_k(t-t_0)^{n-k}}{(n-k)!}\,. \end{aligned}$$

The last one from these equations is a general integral which is, at the same time, the general solution of (5.7).

Now we suppose that the unknown function and some of its derivatives do not appear in (5.4), that is, we have for some $1 \leqslant k < n$

$$F(t, x^{(k)}, x^{(k+1)}, \ldots, x^{(n)}) = 0 \,. \tag{5.8}$$

We introduce the new unknown function $y = x^{(k)}$, then we have

$$F(t, y, y', \ldots, y^{(n-k)}) = 0 \,, \tag{5.9}$$

which is a differential equation of order $n - k < n$. Suppose that we have a general integral of (5.9) in the form

$$\Phi(t, y, C_1, C_2, \ldots, C_{n-k}) = 0 \,, \tag{5.10}$$

which gives the intermediate integral

$$\Phi(t, x^{(k)}, C_1, C_2, \ldots, C_{n-k}) = 0 \tag{5.11}$$

of equation (5.8). If this equation is solvable for $x^{(k)}$, then we have an equation of the type (5.7) which can be solved by successive integration and with the additional constants $C_{n-k+1}, C_{n-k+2}, \ldots, C_n$ we get a general integral of equation (5.8), that is, we obtain the general solution.

On the other hand, if equation (5.11) is not solvable for $x^{(k)}$, then we try to find a parametric equivalent to (5.11) in the form

$$\begin{aligned} t &= \varphi(s) \\ x^{(k)} &= \psi(s) \,. \end{aligned}$$

We have $dx^{(k-1)} = x^{(k)} dt = \psi(s)\varphi'(s)ds$, hence

$$x^{(k-1)} = \int \psi(s)\varphi'(s)ds + C_1 \,.$$

Continuing this process with

$$x^{(k-2)} = \int x^{(k-1)}, \text{ etc.}$$

we arrive at

$$\begin{aligned} t &= \varphi(s) \\ x &= \Phi(s, C_1, C_2, \ldots, C_k) \,, \end{aligned}$$

which is a parametric system of equations. Eliminating the parameter s we obtain a general integral of (5.8).

As an illustration we consider the differential equation

$$e^{y''} + y'' = x \,. \tag{5.12}$$

We parametrize this implicit equation in the following way:

$$x = e^t + t$$
$$y'' = t\,.$$

Then we have

$$dy' = y''dx = t(e^t + 1)dt\,,$$

$$y' = \int t(e^t + 1)dt = (t-1)e^t + \frac{t^2}{2} + C_1\,,$$

further

$$dy = y'dx = \left[(t-1)e^t + \frac{t^2}{2} + C_1\right](e^t + 1)dt\,,$$

and, by integration, we have

$$y = \left(\frac{t}{2} - \frac{3}{4}\right)e^{2t} + \left(\frac{t^2}{2} - 1 + C_1\right)e^t + \frac{t^2}{6} + C_1 t + C_2\,.$$

This latter equation, together with $x = e^t + t$ serves as a parametric form of the general solution of equation (5.12).

Suppose now that the independent variable t does not appear in the equation, that is, we have

$$F(x, x', \ldots, x^{(n)}) = 0\,. \tag{5.13}$$

We introduce the new unknown function $p = x' = \frac{dx}{dt}$ and we consider it as a function of x. Then we have

$$x'' = \frac{dp}{dt} = \frac{dp}{dx}\frac{dx}{dt} = p\frac{dp}{dx}\,,$$

$$x''' = \frac{dx''}{dt} = \frac{d\left(p\frac{dp}{dx}\right)}{dx}\frac{dx}{dt} = p\left[\frac{d^2p}{dx^2} + \left(\frac{dp}{dx}\right)^2\right].$$

Continuing this process we express $x^{(k)}$ in terms of p, $\frac{dp}{dx}, \frac{d^2p}{dx^2}, \ldots, \frac{d^{k-1}p}{dx^{k-1}}$. Then we substitute these expressions for $x'', x''', \ldots, x^{(n)}$ into (5.13) to obtain the differential equation

$$F\left(x, p, \frac{dp}{dx}, \ldots, \frac{d^{n-1}p}{dx^{n-1}}\right) = 0\,, \tag{5.14}$$

which is of order $n-1$. If we have a general integral of this equation in the form

$$\Phi(x, p, C_1, C_2, \ldots, C_{n-1}) = 0\,, \tag{5.15}$$

then, by substituting $p = x'$, we obtain the intermediate integral

$$\Phi(x, x', C_1, C_2, \ldots, C_{n-1}) = 0 \tag{5.16}$$

for equation (5.13) which is a first order differential equation.

5.4 Problems

1. Solve the differential equation:

$$(y''')^2 + x^2 = 1\,.$$

2. Solve the differential equation:

$$y'' = \frac{1}{\sqrt{y}}\,.$$

3. Solve the differential equation:

$$a^3 y''' y'' = (1 + C^2 y'')^{1/2}\,.$$

4. Solve the differential equation, where a is a constant:

$$y''' = \sqrt{1 + (y'')^2}\,.$$

5. Solve the differential equation:

$$y'' - xy''' + (y''')^2 = 0\,.$$

6. Solve the differential equation:

$$y'' - xy''' + (y''')^2 = 0\,.$$

7. Solve the differential equation:

$$yy'' + (y')^2 = y^2 \ln y\,.$$

5.5 Higher order linear differential equations

A complete solution of the linear differential equation (5.3) is called an *integral* of (5.3). The equation

$$x^{(n)} + a_1(t)x^{(n-1)} + \cdots + a_n(t)x = 0 \tag{5.17}$$

is called the *homogeneous equation corresponding to* (5.3).

Theorem 5.5.1. *All integrals of equation* (5.17) *form an n-dimensional linear space over $\mathbb{K}$.*

Proof. The statement follows from Theorem 3.1.2, by the Transition Principle, because the integrals $\varphi_1, \varphi_2, \ldots, \varphi_k$ of (5.17) are linearly independent if and only if the corresponding integrals $\widetilde{\varphi}_1, \widetilde{\varphi}_2, \ldots, \widetilde{\varphi}_k$ are linearly independent. □

Any basis of the space of integrals of (5.17) is called a *fundamental system* of (5.3). For any integrals $\varphi_1, \varphi_2, \ldots, \varphi_n$ of (5.17) the determinant

$$W(t) = \begin{vmatrix} \varphi_1 & \varphi_2 & \cdots & \varphi_n \\ \varphi_1' & \varphi_2' & \cdots & \varphi_n' \\ \cdots & \cdots & \cdots & \cdots \\ \varphi_1^{(n-1)} & \varphi_2^{(n-1)} & \cdots & \varphi_n^{(n-1} \end{vmatrix}$$

is called the *Wronskian* of the functions $\varphi_1, \varphi_2, \ldots, \varphi_n$. This is clearly the same as the Wronskian of the functions $\widetilde{\varphi_1}, \widetilde{\varphi_2}, \ldots, \widetilde{\varphi_k}$ in the sense of Section 3.1. Hence the integrals $\varphi_1, \varphi_2, \ldots, \varphi_n$ form a fundamental system of (5.17) if and only if their Wronskian is different from zero at some point. The theorem corresponding to Liouville's Theorem 3.1.4 also follows from our general principle.

Theorem 5.5.2. *(Liouville) The Wronskian W of any integrals of* (5.17) *satisfies*

$$W(t) = W(t_0) \exp \int_{t_0}^{t} (-a_1(s))ds$$

for each t_0, t in I.

All integrals of equation (5.3) can be determined – like in Section 3.1 – by the method of variation of the constants, if a fundamental system $\varphi_1, \varphi_2, \ldots, \varphi_n$ of (5.3) is known. We try to find a particular solution φ_p of (5.3) in the form

$$\varphi_p(t) = \sum_{k=1}^{n} c_k(t)\varphi_k(t)\,,$$

where we suppose that the unknown functions $c_k : I \to \mathbb{K}$ $(k = 1, 2, \ldots, n)$ satisfy the equations

$$\sum_{k=1}^{n} c_k'(t)\varphi_k^{(i)}(t) = 0, \qquad (i = 1, 2, \ldots, n-2)$$

$$\sum_{k=1}^{n} c_k'(t)\varphi^{(n-1)}(t) = a_0(t)$$

for each t in I. This is a linear system of equations for the unknowns $c_1'(t), c_2'(t), \ldots, c_n'(t)$ which can be solved uniquely, as its determinant is equal to the Wronskian of the functions $\varphi_1, \varphi_2, \ldots, \varphi_n$. It is easy to see that in this case φ_p is an integral of (5.3). As the difference of any two integrals of (5.3) is an integral of (5.17), hence every integral φ of (5.3) has the form

$$\varphi = \varphi_p + c_1\varphi_1 + c_2\varphi_2 + \cdots + c_n\varphi_n$$

with some scalars $c_1, c_2, \ldots, c_n$ in $\mathbb{K}$.

5.6 Problems

1. Find the Wronskian – up to a constant factor – of the differential equation of Legendre:

$$(1-x^2)y'' - 2xy' + n(n+1)y = 0\,,$$

where n is a natural number.

2. Solve the differential equation

$$y'' + \frac{2}{x}y' + y = 0\,,$$

which has the solution $y_p = \frac{\sin x}{x}$.

3. Solve the differential equation

$$y'' \sin^2 x = 2y\,,$$

which has the solution $y_p = \cot x$.

4. Solve the differential equation

$$x^3y''' - 3x^2y'' + 6xy' - 6y = 0\,.$$

5. Solve the differential equation

$$xy''' - y'' + xy' - y = 0\,.$$

6. Solve the differential equation

$$(1-x^2)y''' - xy'' + y' = 0\,.$$

7. Solve the differential equation

$$x^2y'' - 2xy' + 2y = 2x^3\,.$$

8. Solve the differential equation

$$y'' + \frac{x}{1-x}y' - \frac{1}{1-x}y = x - 1\,.$$

9. Solve the differential equation

$$(x^2+2)y''' - 2xy'' + (x^2+2)y' - 2xy = x^4 + 12\,.$$

5.7 Linear differential equations with constant coefficients

If $a_1, a_2, \ldots, a_n$ are constants in (5.17), then (5.17) is called a *linear differential equation with constant coefficients.* It has the general form

$$x^{(n)} + a_1 x^{(n-1)} + \cdots + a_n x = 0\,, \tag{5.18}$$

where the a_i's are constants from $\mathbb{K}$ $(i = 1, 2, \ldots, n)$.

The polynomial

$$P(\lambda) = \lambda^n + a_1\lambda^{n-1} + \cdots + a_n$$

is called the *characteristic polynomial of* (5.18), and its roots are the *characteristic roots.*

In the sequel we shall use the more convenient notation $Dx = x'$. Using this notation (5.18) can be written in the form

$$P(D)x = 0\,.$$

Theorem 5.7.1. *A complex fundamental system of* (5.18) *has the form*

$$\varphi_{k,j}(t) = t^j e^{\lambda_k t}, \qquad (k = 1, 2, \ldots, q;\ j = 0, 1, \ldots n_k - 1)\,,$$

where $\lambda_1, \lambda_2, \ldots, \lambda_q$ *are the different complex roots of the characteristic polynomial of* (5.18) *with multiplicities* $n_1, n_2, \ldots, n_q$*, respectively, where* $n_1 + n_2 + \cdots + n_q = n$.

Proof. As the number of the functions $\varphi_{k,j}$ is equal to n, hence it is enough to show that any complex integral of (5.18) is a linear combination of these functions.

First we note that if $P = Q \cdot R$, where the polynomials Q, R are relatively prime, then every integral φ of the equation $P(D)x = 0$ can be written as the sum of an integral of $Q(D)x = 0$ and of an integral of $R(D)x = 0$. Indeed, by Bezout's Theorem, in this case there are polynomials S, T such that

$$Q \cdot S + R \cdot T = 1\,.$$

Let $\varphi_1 = R(D)T(D)\varphi$ and $\varphi_2 = Q(D)S(D)\varphi$, then we have obviously

$$Q(D)\varphi_1 = R(D)\varphi_2 = 0\,,$$

moreover $\varphi = \varphi_1 + \varphi_2$. This implies that if $\lambda_1, \lambda_2, \ldots, \lambda_q$ are all the different complex roots of the polynomial P with multiplicities $n_1, n_2, \ldots, n_q$, respectively, then φ can be written as a sum of integrals of the equations

$$(D - \lambda_k E)^{n_k} x = 0\,,$$

for $k = 1, 2, \ldots, q$. On the other hand, it is obvious that the equation a

$$(D - \lambda_k E)^{n_k} \varphi_k(t) = 0$$

is equivalent to the equation

$$e^{\lambda_k t} \cdot D^{n_k}(e^{-\lambda_k t} \varphi_k(t)) = 0\,,$$

which holds if and only if the function $t \mapsto e^{-\lambda_k t}\varphi_k(t)$ is a polynomial of degree at most $n_k - 1$. This proves the theorem. □

If equation (5.18) has real coefficients, then a complex valued function is a solution of (5.18) if and only if its real and imaginary parts are solutions, too. In this case it is easy to give a real fundamental system of (5.18). Namely, let $\lambda_1, \lambda_2, \ldots, \lambda_q$ be all the different real roots of the characteristic polynomial, with multiplicities $n_1, n_2, \ldots, n_q$ respectively, further let $\mu_1 \pm i\nu_1$, $\mu_2 \pm i\nu_2, \ldots, \mu_r \pm i\nu_r$ be all the different non-real roots of the characteristic polynomial, with multiplicities $m_1, m_2, \ldots, m_r$ respectively. Obviously, the complex conjugate of every complex root is also a root with the same multiplicity, therefore we have

$$\sum_{j=1}^{q} n_j + 2\sum_{j=1}^{r} m_j = n\,.$$

It is easy to see that in this case the functions

$$t \mapsto t^j e^{\lambda_k t}, \qquad (k = 1, 2, \ldots, q;\ j = 0, 1, \ldots, n_k - 1)$$

and the functions

$$t \mapsto t^j e^{\mu_k t}\cos\nu_k t,\ \ t \mapsto t^j e^{\mu_k t}\sin\nu_k t, \quad (k = 1, 2, \ldots, r;\ j = 0, 1, \ldots, m_k - 1)$$

together form a fundamental system of (5.18).

5.8 Problems

1. Solve the following differential equations:

   i) $y'' + y' - 2y = 0\,,$
   ii) $y'' + 4y' + 3y = 0\,,$
   iii) $y'' - 2y' = 0\,,$
   iv) $2y'' - 5y' + 2y = 0\,,$
   v) $y'' - 4y' + 5y = 0\,,$
   vi) $y'' + 2y' + 10y = 0\,,$
   vii) $y''' - 3y'' + 3y' - y = 0\,,$
   viii) $y''' - y'' - y' + y = 0\,,$

ix) $y^{(iv)} - 5y'' + 4y = 0$,
x) $y^{(v)} + 8y''' + 16y = 0$.

2. Solve the following differential equations:

i) $y'' - 2y' + y = \frac{e^x}{x}$,
ii) $y'' + 3y' + 2y = \frac{1}{e^x+1}$,
iii) $y'' + y = \frac{1}{\sin x}$,
iv) $y''' + 4y = 2\tan x$,
v) $y'' + 2y' + y = 3e^{-x}\sqrt{x+1}$,
vi) $y'' + y = 2\sec^3 x$.

3. Solve the following Cauchy problems:

i) $y'' - 2y' + 2y = 0,\ y(2) = 1,\ y'(2) = -2$,
ii) $y'' + y = 4e^x,\ y(0) = 4,\ y'(0) = -3$,
iii) $y'' - 2y = 2e^x,\ y(1) = -1,\ y'(1) = 0$,
iv) $y'' + 2y' + 2y = xe^{-x},\ y(0) = 0,\ y'(0) = 0$,
v) $y''' - y' = 0,\ y(0) = 3,\ y'(0) = -1,\ y''(0) = 1$,
vi) $y''' - 3y' - 2y = 9e^{2x},\ y(0) = 0,\ y'(0) = -3,\ y''(0) = 3$,
vii) $y^{(iv)} + y'' = 2\cos x,\ y(0) = -2,\ y'(0) = 1,\ y''(0) = 0,\ y'''(0) = 0$.

5.9 Decreasing the order of linear homogeneous equations

We have seen above that the solution of the homogeneous linear differential equations is of utmost importance in the theory of linear equations. In the constant coefficient case this problem is reduced to a pure algebraic job: finding the roots of polynomials. Unfortunately, in the case of variable coefficients, there are no general methods for the complete description of the solution space of the homogeneous equation. Nevertheless, in some special cases there are possibilities to reduce the problem by decreasing the order of the equation. In this section we exhibit some of these methods.

Suppose first that we are given the second order homogeneous equation

$$y'' + a(x)y' + b(x)y = 0, \tag{5.19}$$

where $a, b : I \to \mathbb{R}$ are continuous on the open interval $I \subseteq \mathbb{R}$. Suppose that $y_1 \neq 0$ is a solution of (5.19). For each solution y we have that the Wronskian of y and y_1 is the following:

$$W(x) = \begin{vmatrix} y & y_1(x) \\ y' & y_1'(x) \end{vmatrix} = y_1'(x)y - y_1(x)y'.$$

If y and y_1 are linearly independent, then $W(x) \neq 0$. This means, by Theorem 5.5.2, that each solution y_2 of the first order linear differential equation

$$y_1(x)y' - y_1'(x)y = \exp \int -a(x)dx$$

gives a solution of (5.19) which is linearly independent of y_1. Hence y_1, y_2 form a fundamental system of (5.19), and the general solution can be described.

We consider the following example:

$$(1-x^2)y'' - 2xy' + 2y = 0\,.$$

Let us try to find a polynomial solution of this equation. If we try with $y(x) = x^n$, then, by substitution, we immediately get the particular solution $y_1(x) = x$. By the above consideration we have to solve the differential equation

$$xy' - y = \exp \frac{2x}{1-x^2} = \frac{1}{1-x^2}\,,$$

or, what is the same

$$y' - \frac{1}{x}y = \frac{1}{x(1-x^2)}\,.$$

We do not need the general solution of this equation, just a particular solution is quite enough. We obtain it using the variation of the constants:

$$y = \frac{1}{2}x \ln \frac{1+x}{1-x} - 1\,.$$

Hence a fundamental system of our original equation is

$$x, \frac{1}{2}x \ln \frac{1+x}{1-x} - 1\,,$$

and the general solution is

$$y = C_1 x + C_2 \frac{1}{2}x \ln \frac{1+x}{1-x} - 1\,.$$

This means that knowing a non-zero solution of a second order linear homogeneous differential equation we can find another one, so that the two solutions form a fundamental system. Roughly speaking, we can "decrease the order" of the equation by one.

Now we show that knowing a non-zero solution of the n-th order linear homogeneous differential equation we can decrease the order of the equation by one.

Suppose that y_1 is a solution of the n-th order linear homogeneous differential equation (5.17), and we introduce the new unknown function z by $x = y_1 \cdot z$. First we compute the derivatives of z:

$$
\begin{aligned}
z' &= y_1 z' + y_1' z \\
z'' &= y_1 z'' + 2y_1' z' + y_1'' z \\
&\dots \\
&\dots \\
&\dots \\
z^{(n)} &= y_1 z^{(n)} + \binom{n}{1} y_1' z^{(n-1)} + \binom{n}{2} y_1'' z^{(n-2)} + \cdots + y_1^{(n)} z\,.
\end{aligned}
$$

Substitution into (5.17) gives

$$y_1 z^{(n)} + \left[\binom{n}{1} y_1' + a_1 y_1\right] z^{(n-1)} + \cdots + [y^{(n)} + a_1 y^{(n-1)} + \cdots + a_{n-1} y' + a_n y] z = 0\,.$$

This is still an n-th order equation for z, however, the coefficient of z is zero, as y_1 is a solution of (5.17). We apply the substitution $u = z' = \frac{d}{dt}\left(\frac{x}{y_1}\right)$, then the above equation is of order $n-1$ for u:

$$u^{(n-1)} + b_1 u^{(n-2)} + \cdots + b_{n-1} u = 0\,. \tag{5.20}$$

If it has the fundamental system $u_1, u_2, \ldots, u_{n-1}$, then a fundamental system of the equation for z will be $1, \int u_1, \int u_2, \ldots, \int u_{n-1}$, and a fundamental system of the equation for x is

$$y_1, y_2 = y_1, \int u_1, y_3 = y_1 \int u_2, \ldots, y_n = y_1 \int u_{n-1}\,.$$

Indeed, it is easy to check that the functions $y_1, y_2, \ldots, y_n$ are linearly independent.

We can see easily that knowing r linearly independent solutions $y_1, y_2, \ldots, y_r$ of (5.17), with $1 \leqslant r < n$ we can decrease the order of the equation by r. Indeed, the functions

$$u_1 = \left(\frac{y_2}{y_1}\right), u_2 = \left(\frac{y_3}{y_1}\right), \ldots, u_{r-1} = \left(\frac{y_r}{y_1}\right)$$

are linearly independent solutions of (5.20). Now we can repeat our argument for equation (5.20) with the substitution $v = \frac{u}{u_1}$ and continuing this process finally we arrive at an equation of order $n - r$.

5.10 Problems

1. Find polynomial solutions and the general solution of the equation

$$x^3y''' - 3x^2y'' + 6xy' - 6y = 0\,.$$

2. Find polynomial solutions and the general solution of the equation

$$xy''' - y'' + xy' - y = 0\,.$$

3. Find polynomial solutions and the general solution of the equation

$$(1 - x^2)y''' - xy'' + y' = 0\,.$$

5.11 Euler differential equations

In this section we present a class of differential equations with variable coefficients which can be transformed into constant coefficient equations by a simple substitution. We consider the differential equation

$$x^n y^{(n)} + a_1 x^{n-1} y^{(n-1)} + a_2 x^{n-2} y^{(n-2)} + \cdots + a_{n-1} xy' + a_n y = 0 \quad (5.21)$$

on the positive reals, where $a_1, a_2, \ldots, a_n$ are constants. This equation is called *Euler differential equation.* We apply the substitution $t = \ln x$, then the equation will be transformed into a linear differential equation of order n with constant coefficients. Indeed, we have $x = e^t$, and

$$\frac{d^k y}{dx^k} = e^{-kt}\Big(\frac{d^k y}{dt^k} + \alpha_1 \frac{d^{k-1} y}{dx^{k-1}} + \cdots + \alpha_{k-1}\frac{dy}{dt}\Big) \quad (5.22)$$

for $k = 1, 2, \ldots, n$, with some constants $\alpha_1, \alpha_2, \ldots, \alpha_{k-1}$ which can be proved easily, by induction. It follows that, by multiplication with $x^k = e^{kt}$, we obtain an n-th order linear differential equation with constant coefficients.

We consider the following example:

$$x^2y'' + 3xy' + y = 0\,.$$

Substituting $t = \ln x$ we get the equation

$$\frac{d^2y}{dt^2} + 2\frac{dy}{dt} + y = 0\,,$$

where $y = y(t)$ is a function of t. The characteristic equation is

$$(\lambda + 1)^2 = \lambda^2 + 2\lambda + 1 = 0\,,$$

and the characteristic roots are $\lambda_{1,2} = -1$. It follows that a fundamental system of the equation is $\{e^{-t}, te^{-t}\}$, hence the general solution is $y(t) = e^{-t}(C_1 + C_2 t)$. Finally, the general solution of the original equation is

$$y(x) = \frac{1}{x}(C_1 + C_2 \ln x)\,.$$

We remark that similar idea can be applied in the case of the following, slightly more general equation:

$$(ax+b)^n y^{(n)} + a_1(ax+b)^{n-1} y^{(n-1)} \tag{5.23}$$

$$+ a_2(ax+b)^{n-2} y^{(n-2)} + \cdots + a_{n-1}(ax+b)y' + a_n y = 0\,.$$

Here we apply the substitution $ax + b = e^t$.

5.12 Problems

1. Find the general solution of the equation

$$x^2y'' - 4xy' + 6y = x\,.$$

2. Find the general solution of the equation

$$x^2y'' - xy' + 2y = x \ln x\,.$$

3. Find the general solution of the equation

$$x^2y'' - 2y = x^2 + \frac{1}{x}\,.$$

4. Find the general solution of the equation

$$x^3y''' - x^2y'' + 2xy' - 2y = x^3 + 3x\,.$$

5. Find the general solution of the equation

$$(1+x)^2y'' + (1+x)y' + y = 4\cos\ln(1+x)\,.$$

5.13 Exponential polynomials

In this section we exhibit a method which can be used to find particular solution for a given n-th order inhomogeneous linear differential equation with constant coefficients, if the right side of the equation belongs to a special class of functions, to the class of the so-called exponential polynomials. In this section we shall consider complex valued functions and differential equations with complex constant coefficients.

We consider the differential equation

$$x^{(n)} + a_1 x^{(n-1)} + a_2 x^{(n-2)} + \cdots + a_{n-1} x' + a_n x = g(x)\,, \tag{5.24}$$

where $a_, a_2, \ldots, a_n$ are complex numbers and $g : \mathbb{R} \to \mathbb{C}$ is a continuous function. The left side of this equation will be denoted by $L[x]$, and the characteristic polynomial of (5.24) will be denoted by $p(\lambda)$. By the results of Section 5.7, equation (5.24) has a fundamental system consisting of functions of the form $t \mapsto P(t)e^{\alpha t}$, where $P : \mathbb{R} \to \mathbb{C}$ is a polynomial and α is a complex number. Such functions will be called *exponential monomials*, and linear combinations of exponential monomials will be called *exponential polynomials*. Exponential functions are obviously exponential monomials. We say that the non-zero exponential monomial $t \mapsto P(t)e^{\alpha t}$ *corresponds* to the exponential $t \mapsto e^{\alpha t}$. Each non-zero exponential polynomial g has a representation in the form

$$g(t) = \sum_{k=1}^{m} P_k(t) e^{\alpha_k t} \tag{5.25}$$

with some non-zero polynomials $P_1, P_2, \ldots, P_m$ and different complex numbers $\alpha_1, \alpha_2 \ldots, \alpha_m$. It is easy to prove that this representation is unique which we call the *canonical representation* of g. We consider $g = 0$ the canonical representation of the zero exponential polynomial. It follows that non-zero exponential monomials corresponding to different exponentials are linearly independent.

Suppose that the function g in (5.24) is an exponential polynomial of the form (5.25), and we are looking for a particular solution of (5.24). By the linearity of the equation, it is obvious that we can solve this problem assuming that we can solve it, whenever g is an exponential monomial. Consequently, we may assume that

$$g(t) = P(t)e^{\alpha t}\,, \tag{5.26}$$

where P is a non-zero polynomial of degree $m \geqslant 0$ of the form

$$P(t) = p_m t^m + p_{m-1} t^{m-1} + \cdots + p_1 t + p_0 \tag{5.27}$$

with some complex numbers p_j, $j = 0, 1, \ldots, m$, $p_m \neq 0$, and α is a root of the characteristic polynomial p with multiplicity $r \geqslant 0$. Observe that our condition on r allows that α is not a root of p: this is the case $r = 0$. In other words, our condition on r means that

$$p(\alpha) = p'(\alpha) = \cdots = p^{(r-1)}(\alpha) = 0, \;\; p^{(r)}(\alpha) \neq 0\,.$$

We show that in this case there exists a solution of (5.24) of the form

$$x(t) = t^r Q(t) e^{\alpha t} = t^r (q_m t^m + t^{m-1} t^{m-1} + \cdots + q_1 t + q_0) e^{\alpha t}$$

with some complex numbers $q_0, q_1, \ldots, q_m$. Indeed, if we substitute x into (5.24), then we have

$$e^{-\alpha t} L[t^r Q(t) e^{\alpha t}] = P(t)\,.$$

Here the left hand side is

$$e^{-\alpha t} L[t^r Q(t) e^{\alpha t}] = \sum_{j=0}^{m} q_j \sum_{i=0}^{j} \binom{r+j}{r+i} t^{j-i} p^{(r+i)}(\alpha)\,,$$

as it is easy to check, by differentiation. Comparing the coefficients of the polynomials on both sides we obtain:

$$p_{m-l} = \sum_{i=0}^{l} \binom{m+r+i-l}{r+i} p^{(r+i)}(\alpha) q_{m-l+i}$$

for $l = 0, 1, \ldots, m$. This is a linear system of equations for the unknowns $q_0, q_1, \ldots, q_m$, whose fundamental matrix is diagonal and its determinant is

$$\binom{m+r}{r}\binom{m+r-}{r}\binom{m+r-2}{r}\cdots\binom{r}{r}[p^{(r)}(\alpha)]^{m+1} \neq 0\,,$$

which has a unique solution, by Cramer's Rule. Using the solution $q_0, q_1, \ldots, q_m$ as coefficients of Q the corresponding exponential monomial $x \mapsto t^r Q(t) e^{\alpha t}$ is a particular solution of (5.24). It follows that if g in (5.24) is an exponential polynomial, then (5.24) can be solved without integration, and the general solution is a family of exponential polynomials.

We illustrate this method by the following example: we solve the differential equation

$$y''' + y'' = x^2 + 1 + 3xe^x\,. \tag{5.28}$$

The roots of the characteristic polynomial $p(\lambda) = \lambda^3 + \lambda^2$ are $\lambda_1 = \lambda_2 = 0$ and $\lambda_3 = -1$. Our problem splits into the following equations:

$$y''' + y'' = x^2 + 1\,, \tag{5.29}$$

and

$$y''' + y'' = 3xe^x\,. \tag{5.30}$$

In the first case the right side corresponds to the exponential with exponent $\alpha = 0$ which is a double root of the characteristic polynomial, accordingly we look for a particular solution of (5.29) in the form

$$y_1(x) = x^2 Q(x) = q_2 x^4 + q_1 x^3 + q_0 x^2\,.$$

Substitution into (5.29) gives

$$24q_2x + 6q_1 + 12q_2x^2 + 6q_1x + 2q_0 = x^2 + 1\,.$$

Comparing the coefficients of x we get the solution of (5.29):

$$y_1(x) = \frac{1}{12}x^4 - \frac{1}{3}x^3 + \frac{3}{2}x^2\,.$$

In the case of the second equation (5.30) the right side corresponds to the exponential with exponent $\alpha = 1$ which is not a root of the characteristic polynomial, hence we look for a particular solution y_2 in the form

$$y_2(x) = Q(x)e^x = (q_1x + q_0)e^x\,.$$

Similarly like above, performing the computations we obtain the particular solution of (5.30) in the form

$$y_2(x) = \Big(\frac{3}{2}x - \frac{15}{4}\Big)e^x\,.$$

Finally, we have the general solution of (5.28) as

$$y(x) = C_1e^{-x} + C_2 + C_3x + \frac{1}{12}x^4 - \frac{1}{3}x^3 + \frac{3}{2}x^2 + \Big(\frac{3}{2}x - \frac{15}{4}\Big)e^x\,.$$

5.14 Problems

Find the general solution of the following differential equations:

1. $y'' - 4y' + 4 = x^2$
2. $y'' - 6y' + 8y = e^x + e^{2x}$
3. $y''' + y'' + y' + y = xe^x$
4. $y^{(iv)} - 4y''' + 6y'' - 4y' + y = (x+1)e^x$
5. $y'' + 4y = x\sin 2x$

5.15 Boundary value problems

Besides initial value problems there is another type of problems concerning differential equations which plays a central role in the theory and in the applications, as well. We devote this section to a class of this type problems: the boundary value problems. The modern theory of boundary value problems depends on deep results of functional analysis which are out of the scope of this volume. However, we consider here a rather special case of this problem and we give an insight into the possible solution methods.

Let $[a,b]$ be a finite closed interval in $\mathbb{R}$ and let $p_0, p_1, p_2, r : [a,b] \to \mathbb{R}$ be continuous functions with $p_0 \neq 0$ for each $a \leqslant t \leqslant b$. Let $\alpha, \beta, \gamma, \delta$ be real numbers with $|\alpha| + |\beta| \neq 0$ and $|\gamma| + |\delta| \neq 0$. Then we consider the problem

$$p_0(t)x'' + p_1(t)x' + p_2(t)x = r(t) \tag{5.31}$$

$$\alpha x'(a) + \beta x(a) = 0 \tag{5.32}$$

$$\gamma x'(b) + \delta x(b) = 0\,. \tag{5.33}$$

This problem is called *boundary value problem* for the second order differential equation (5.31). In this problem, equations (5.32) and (5.33) are called *boundary conditions*. The concept of the *solution* of this problem is quite obvious: a solution of the boundary value problem (5.31)-(5.33) is a continuous function $\varphi : [a,b] \to \mathbb{R}$ which is twice differentiable on $]a,b[$, and its first and second one sided derivatives exist at the endpoints such that they are continuous on $[a,b]$, too. For the sake of simplicity we call such functions *twice continuously differentiable* on $[a,b]$.

Together with the above boundary value problem we consider the following one:

$$p_0(t)x'' + p_1(t)x' + p_2(t)x = 0 \tag{5.34}$$

$$\alpha x'(a) + \beta x(a) = 0 \tag{5.35}$$

$$\gamma x'(b) + \delta x(b) = 0\,, \tag{5.36}$$

which is called the corresponding *homogeneous boundary value problem.* For the left hand side of (5.31) and (5.34) we use the notation

$$L[x] = p_0(t)x'' + p_1(t)x' + p_2(t)x\,.$$

In contrast with initial value problems boundary value problems may not have any solution, or they may have more than one, possibly infinitely many solutions. As an illustration we consider the following example. Let $p_0 = p_2 = r = 1$, and $p_1 = 0$. The general solution of the second order homogeneous linear differential equation (5.31) on any interval has the form

$$x(t) = A\cos t + B\sin t + 1$$

with some real constants A, B. If we let $[a,b] = [0, 2\pi]$, $\alpha = \gamma = 0$, and $\beta = \delta = 1$, then we have $A = -1$ and B is arbitrary, hence in this case any function of the form

$$x(t) = -\cos t + B\sin t + 1$$

is a solution of our problem with arbitrary real constant B. Consequently, in this case we have infinitely many solutions. However, if we let $[a, b] = [0, 2\pi]$, $\alpha = \delta = 0$, and $\beta = \gamma = 1$, then we have $A = -1$ and $B = 0$, hence the only solution of the problem is

$$x(t) = -\cos t + 1\,,$$

which is the unique solution. Finally, if we let $[a, b] = [0, \pi]$, $\alpha = \gamma = 0$, and $\beta = \delta = 1$, then there is no solution.

Concerning the existence and uniqueness questions of the boundary value problem (5.31)-(5.33) we have the following theorem.

Theorem 5.15.1. *(Alternative Theorem) Given the boundary value problem* (5.31)-(5.33) *we have the following two possibilities:*

1. *For every right side and for every boundary conditions the boundary value problem* (5.31)-(5.33) *has a unique solution.*
2. *The homogeneous boundary value problem* (5.34)-(5.36) *has infinitely many solutions, for some non-zero right side the problem* (5.31)-(5.33) *has infinitely many solutions, and for some right side the problem* (5.31)-(5.33) *has no solution.*

Proof. It is known that the general solution of (5.31) has the form

$$x = c_1 x_1 + c_2 x_2 + x_p\,,$$

where x_1, x_2 are linearly independent solutions of the homogeneous equation (5.34), x_p is a particular solution of (5.31), and c_1, c_2 are arbitrary constants. Substituting the boundary conditions we obtain a system of two linear equations for the unknowns c_1, c_2. We have two possibilities. If the determinant of this system is non-zero, then it has a unique solution for every right side and the boundary value problem (5.31)-(5.33) has a unique solution. In the other case the determinant of the system is zero, the homogeneous equation system has infinitely many solutions c_1, c_2, and the inhomogeneous system cannot be solved for every right side. If it has a solution for some right side, then it has infinitely many solutions for this right side, because we can add to the solution any solution of the homogeneous boundary value problem. Otherwise it has no solution which completes the proof of the theorem. □

We consider the following example: find the smallest positive value of b for which the boundary value problem

$$y'' + b^2 y = 0,\ y(0) = 5,\ y(1) = -5 \qquad (5.37)$$

has no solution. The general solution of the differential equation has the form

$$y(x) = Ay_1(x) + By_2(x)\,, \tag{5.38}$$

where y_1, y_2 are linearly independent solutions and A, B are arbitrary real numbers. Substituting the boundary conditions we obtain

$$\begin{aligned} Ay_1(0) + By_2(0) &= 5 \\ Ay_1(1) + By_2(1) &= -5\,. \end{aligned} \tag{5.39}$$

If the determinant

$$d = \begin{vmatrix} y_1(0) & y_2(0) \\ y_1(1) & y_2(1) \end{vmatrix}$$

is non-zero, then the system of linear equations (5.39) has a unique solution. Hence if our boundary value problem has no solution, then $d = 0$. It follows that the system of homogeneous equations

$$\begin{aligned} Cy_1(0) + Dy_2(0) &= 0 \\ Cy_1(1) + Dy_2(1) &= 0 \end{aligned} \tag{5.40}$$

has non-trivial solution which implies that the boundary value problem

$$y'' + b^2 y = 0,\ y(0) = 0,\ y(1) = 0$$

has non-zero solution. The non-zero solutions y of this second order homogeneous linear equation which satisfy $y(0) = 0$ have the form $y(x) = c \sin bx$ with $c \neq 0$, and, by $y(1) = 0$, we have $\sin b = 0$ which gives $b = k\pi$ with some integer k. As $b > 0$, we have $k = 1, 2, \ldots$. For $k = 1$ we have $b = \pi$ and the general solution of the differential equation in (5.37) is

$$y(x) = A \cos \pi x + B \sin \pi x$$

with some real numbers A, B. By the boundary conditions we have $y(0) = A$, $y(1) = -A$, hence the function

$$y(x) = 5 \cos \pi x + B \sin \pi x$$

is a solution of (5.37). For $k = 2$ we have $b = 2\pi$ and, by the boundary conditions, we have from (5.38): $5 = y(0) = A$, and $-5 = y(1) = A$ which is impossible. Hence the answer is $b = 2\pi$.

From this example it is clear that existence and uniqueness depends heavily on the basic interval and on the constants occur in the boundary conditions. In this section we do not study existence and uniqueness questions for the problem (5.31)-(5.33) in general, but we restrict ourselves to some basic results and methods which can be used to find solutions of boundary value problems of the above type.

The function $G : [a,b] \times]a,b[\to \mathbb{R}$ is called the *Green function* of the boundary value problem (5.31)-(5.33), if the following conditions are satisfied:

1. the function $t \mapsto G(t,s)$ is a solution of (5.31) for every $t \neq s$ in $[a,b]$, whenever s is in $]a,b[$,
2. the function $t \mapsto G(t,s)$ satisfies the boundary conditions (5.32)-(5.33), whenever s is in $]a,b[$,
3. we have

$$\lim_{t \to s-0} G(t,s) = \lim_{t \to s+0} G(t,s), \tag{5.41}$$

$$\lim_{t \to s-0} \frac{d}{dt} G(t,s) + \frac{1}{p_0(s)} = \lim_{t \to s+0} \frac{d}{dt} G(t,s),$$

whenever s is in $]a,b[$.

In other words, the Green function is a solution of the given boundary value problem in its first variable for $t \neq s$, it is continuous at s, and its derivative with respect to the first variable has a discontinuity with jump $\frac{1}{p_0(s)}$ at s. The following theorem is about the existence and the computation of the Green function.

Theorem 5.15.2. *Suppose that p_0, p_1, p_2 are continuous in* (5.31)*, further $p_0 \neq 0$ on $[a,b]$ and $r \equiv 0$. If the only solution of the boundary value problem* (5.31)-(5.33) *is the zero function, then the Green function of the boundary value problem exists and it has the form*

$$G(t,s) = \begin{cases} c y_1(t) & \text{if } a \leqslant t \leqslant s \\ d y_2(t) & \text{if } s \leqslant t \leqslant b, \end{cases} \tag{5.42}$$

where y_1, respectively y_2 are non-zero solutions of (5.31) *on $[a,b]$, satisfying the boundary conditions* (5.32)*, respectively* (5.33)*, and c,d are functions of s such that G satisfies the conditions* (5.41)*:*

$$c y_1(s) = d y_2(s), \; b y_1'(s) + \frac{1}{p_0(s)} = d y_2'(s). \tag{5.43}$$

Proof. Let y_1, respectively y_2 be the solution of (5.31) with $r = 0$ satisfying the initial conditions $y_1(a) = \alpha$, $y_1'(a) = -\beta$, respectively $y_2(b) = \gamma$, $y_2'(b) = -\delta$. If y_1, y_2 are linearly dependent, then we have $y_1 = \lambda y_2$. As $|\gamma| + |\delta| \neq 0$ it follows $y_1 \neq 0$ and the non-zero function y_1 satisfies both boundary conditions in (5.31) which is impossible, by assumption. Hence y_1, y_2 are linearly independent. Consequently, every solution of (5.31) with $r = 0$ has the form $y = c_1 y_1 + c_2 y_2$ with some constants c_1, c_2. By the first two properties of the Green function G must have the given form. The third condition (5.41) implies (5.43). The determinant

$$W(s) = \begin{vmatrix} y_1(s) & -y_2(s) \\ -y_1'(s) & y_2'(s) \end{vmatrix}$$

is non-zero, as y_1, y_2 are linearly independent, hence the equation system (5.43) can be solved for c, d, and the function G given in (5.42) using c, d obtained satisfies the requirements of the Green function. □

We note that, by the proof of the previous theorem, the Green function of the boundary value problem (5.31)-(5.33) is unique.

Theorem 5.15.3. *Suppose that the conditions of Theorem* 5.15.2 *are satisfied with some continuous function r. Then the solution x of the boundary value problem* (5.31)-(5.33) *can be given as*

$$x(t) = \int_a^b G(t,s) r(s)\, ds\,. \tag{5.44}$$

Proof. The proof of this statement is a simple calculation based on differentiation of parametric integrals. The details are left to the reader. □

Sometimes we consider parametric boundary value problems of the following type:

$$p_0(t)x'' + p_1(t)x' + p_2(t)x = \lambda x \tag{5.45}$$

$$\alpha x'(a) + \beta x(a) = 0 \tag{5.46}$$

$$\gamma x'(b) + \delta x(b) = 0 \tag{5.47}$$

with the same conditions as in (5.31)-(5.33). Here λ is a real number which is called the *parameter*. The number λ is called an *eigenvalue* of the problem (5.45)-(5.47), if (5.45)-(5.47) has a non-zero solution. In this case the non-zero solutions are called *eigenfunctions* corresponding to the given eigenvalue.

5.16 Problems

1. Find the smallest positive p such that the boundary value problem
$$y'' + py = 0, \quad y(0) = 1,\ y(1) = 2$$
has no solution.
2. Find the largest real number a such that for each p in the interval $]1, a[$ the boundary value problem
$$y'' + 2y' + py = 0, \quad y(0) = 2,\ y(\pi) = 3$$
has a solution.
3. Find the Green function of the following boundary value problems:
   i) $y'' = f(x),\ \ y(0) = 0,\ y(1) = 0$
   ii) $y'' + y = f(x),\ \ y'(0) = 0,\ y(\pi) = 0$
   iii) $y'' + y' = f(x),\ \ y(0) = 0,\ y'(1) = 0$
   iv) $y'' - y = f(x),\ \ y'(0) = 0,\ y'(2) + y(2) = 0$
   v) $x^2y'' + 2xy' = f(x),\ \ y(1) = 0,\ y'(3) = 0$
   vi) $xy'' - y' = f(x),\ \ y(0) = 0,\ y(\pi) = 0$
   vii) $x^2y'' - 2y = f(x),\ \ y(1) = 0,\ y(2) + 2y'(2) = 0$
4. Find the eigenvalues and the corresponding eigenfunctions of the following boundary value problems:
   i) $y'' = \lambda y,\ \ y(0) = 0,\ y(l) = 0$
   ii) $y'' = \lambda y,\ \ y'(0) = 0,\ y'(l) = 0$
   iii) $y'' = \lambda y,\ \ y(0) = 0,\ y'(l) = 0$
   iv) $x^2y'' = \lambda y,\ \ y(1) = 0,\ y(a) = 0\ \ (a > 1)$
5. Solve the following boundary value problems:
   i) $y'' - y = 2x,\ \ y(0) = 0,\ y(1) = -1$
   ii) $y'' + y' = 1,\ \ y'(0) = 0,\ y(1) = 1$
   iii) $y'' - y' = 0,\ \ y(0) = -1,\ y'(1) - y(1) = 2$
   iv) $y'' + y = 1,\ \ y(0) = 0,\ y(\frac{\pi}{2}) = 0$
   v) $y'' + y = 1,\ \ y(0) = 0,\ y(\pi) = 0$
   vi) $y'' + y = 2x - \pi,\ \ y(0) = 0,\ y(\pi) = 0$

5.17 Power series solutions

In this section we discuss the possibility to find solutions of second order linear differential equations using power series. We note that the ideas in this section can be generalized for higher order linear and even for nonlinear differential equations. However, here we consider the second order case only. The interested reader should consult with the vast literature on differential equations.

We consider the following differential equation

$$y'' + p(x)y' + q(x)y = r(x)\,, \tag{5.48}$$

and the corresponding homogeneous equation

$$y'' + p(x)y' + q(x)y = 0\,, \tag{5.49}$$

where p, q, r are continuous functions on the non-empty open interval $I \subseteq \mathbb{R}$.

We recall that if x_0 is a point in I, then the function $f : I \to \mathbb{R}$ is called *analytic* at x_0, if its Taylor series at x_0 has a positive convergence radius, and it is equal to the sum of this Taylor series on the corresponding convergence interval. We note that in this case f is obviously infinitely differentiable, each derivative is analytic at x_0 with the same convergence interval, and the derivatives can be computed by term-wise differentiation of the corresponding Taylor series. The point x_0 is called an *ordinary point* of (5.48), if both p and q are analytic at x_0. Otherwise it is called a *singular point.* We have the following theorem.

Theorem 5.17.1. *Let x_0 be an ordinary point of* (5.48) *in I. Then there is an open interval J in I containing x_0 such that the general solution of* (5.49) *in J can be given in the form*

$$y = c_1 y_1 + c_2 y_2\,, \tag{5.50}$$

where $y_1, y_2 : J \to \mathbb{R}$ are analytic at x_0, y_1 and y_2 are linearly independent solutions of (5.49), *and c_1, c_2 are arbitrary constants.*

The statement of this theorem can be derived form our earlier results on linear differential equations in Section 5.5, however, the proof of the analyticity needs some additional tools. Without going into the details here we show instead how this can be used for the description of the solutions of (5.48). By assumption we have

$$p(x) = \sum_{n=0}^{\infty} p_n (x - x_0)^n\,, \quad q(x) = \sum_{n=0}^{\infty} q_n (x - x_0)^n\,, \tag{5.51}$$

for $|x - x_0| < r$, where $r > 0$ is the convergence radius of the power series.

By the theorem, we have that the solution has the power series representation

$$y(x) = \sum_{n=0}^{\infty} a_n (x - x_0)^n \tag{5.52}$$

for $|x - x_0| < r$. Then we have

$$y'(x) = \sum_{n=0}^{\infty} (n + 1) a_{n+1} (x - x_0)^n \tag{5.53}$$

and

$$y''(x) = \sum_{n=0}^{\infty} (n + 2)(n + 1) a_{n+2} (x - x_0)^n \tag{5.54}$$

for $|x - x_0| < r$. Substituting the expressions for p, q, y, y', y'' into (5.48) and collecting the terms containing $(x - x_0)^n$ for $n = 0, 1, \ldots$ we obtain a power series which is convergent on the non-empty open interval J, and its sum is zero. It follows, by the uniqueness of power series representation, that the coefficients of the terms $(x - x_0)^n$ are equal to zero for $n = 0, 1, \ldots$. This results in a system of infinitely many linear equations, from which we can derive a recursive equation for the unknown coefficients a_n, $n = 0, 1, \ldots$. Solving this recursion, if possible, we obtain the power series representation of the solution.

In the case of the inhomogeneous equation we have the similar result.

Theorem 5.17.2. *Let x_0 be an ordinary point of* (5.48) *in I and suppose that r is analytic at x_0. Then there is an open interval J in I containing x_0 such that the general solution of* (5.49) *in J can be given in the form*

$$y = c_1 y_1 + c_2 y_2 + y_0 \,, \tag{5.55}$$

where $y_1, y_2, y_0 : J \to \mathbb{R}$ are analytic at x_0, y_1 and y_2 are linearly independent solutions of (5.49)*, y_0 is a solution of* (5.48)*, and c_1, c_2 are arbitrary constants.*

In this case we can follow the above method using the power series representation of r.

We consider the following application:

$$y'' - xy' + 2y = 0 \,. \tag{5.56}$$

Here we take $x_0 = 0$, then $p(x) = -x$ and $q(x) = 2$ are analytic at x_0, hence x_0 is an ordinary point. This means that we look for a solution around $x_0 = 0$ in the form of the power series

$$y(x) = \sum_{n=0}^{\infty} a_n x^n .$$

We have

$$y'(x) = \sum_{n=0}^{\infty} (n+1)a_{n+1}x^n, \quad y''(x) = \sum_{n=0}^{\infty} (n+2)(n+1)a_{n+2}x^n .$$

Substituting into (5.56) we have

$$\sum_{n=0}^{\infty} [(n+2)(n+1)a_{n+2}x^n - \sum_{n=1}^{\infty} na_n x^n + \sum_{n=0}^{\infty} 2a_n x^n = 0 .$$

By the uniqueness of power series representation, we compare the coefficients of x^n on the two sides and we have

$$a_2 + a_0 = 0, \quad (n+2)(n+1)a_{n+2} - (n-2)a_n = 0$$

for $n = 1, 2, \ldots$. This is a second order recursion for the unknown coefficients a_n. Taking a_0, a_1 arbitrarily we can determine the values $a_2, a_3, \ldots$ uniquely from the formula

$$a_{n+2} = \frac{n-2}{(n+2)(n+1)} a_n, \quad n = 1, 2, \ldots$$

The above method works in the case of ordinary points. If x_0 is a singular point, then we call it a *regular singular point*, if the functions $x \mapsto (x - x_0)p(x)$ and $x \mapsto (x - x_0)^2 q(x)$ are analytic at x_0. For regular singular point we have the following theorem.

Theorem 5.17.3. *(Frobenius) Let x_0 be a regular singular point of (5.48) in I. Then there is a positive number $R > 0$, and there exists a solution of (5.48) on the interval $[x_0, x_0 + R[$ of the form*

$$y(x) = (x - x_0)^\lambda \sum_{n=0}^{\infty} a_n (x - x_0)^n \tag{5.57}$$

with some constant λ.

The method of Frobenius is similar to the case of an ordinary point: we substitute the expression of y in (5.57) into (5.49), and then, by equating the coefficients of the different powers of x on the two sides of the equation we obtain recursive formulas for the coefficients a_n. A quadratic equation arises for λ, when the coefficient of $x^0 = 1$ is set to zero and a_0 is left arbitrary. This quadratic equation is called *indicial equation.* It may have real or complex roots. If $\lambda = a+ib$ is a complex root with a, b real numbers, then its conjugate $\overline{\lambda} = a - ib$ is the other root, and the complex solutions, including $x^{a \pm ib} = x^a e^{\pm ib \ln x}$, can be combined to get linearly independent real solutions. For simplicity, here we consider the case of real roots, only. If λ is taken as the larger indicial root, $\lambda = \lambda_1 \geqslant \lambda_2$, then the method of Frobenius always yields a solution y_1 of the form (5.57). To get another solution y_2 such that y_1, y_2 are linearly independent we consider three cases.

1. Case 1: $\lambda_1 - \lambda_2$ is not an integer. Then we take

$$y_2(x) = y(x) = (x - x_0)^{\lambda_2} \sum_{n=0}^{\infty} a_n (x - x_0)^n \,,$$

and we follow the method exactly in the same way like in the case of λ_1.

2. Case 2: $\lambda_1 = \lambda_2$. Then we look for y_2 in the form

$$y_2(x) = y_1(x) \ln(x - x_0) + (x - x_0)^{\lambda_2} \sum_{n=0}^{\infty} b_n (x - x_0)^n \,.$$

We can generate this solution in te following way: solve the recurrence formula for a_n in terms of λ and a_0, then substitute the a_n's into the expression of y. We obtain a function $y = y(\lambda, x)$ with the arbitrary parameter a_0. Then we take

$$y_2(x) = \left. \frac{\partial y(\lambda, x)}{\partial \lambda} \right|_{\lambda = \lambda_1} .$$

3. Case 3: $\lambda_1 - \lambda_2$ is a positive integer. Then we look y_2 in the form

$$y_2(x) = d_{-1} y_1(x) \ln(x - x_0) + (x - x_0)^{\lambda_2} \sum_{n=0}^{\infty} d_n (x - x_0)^n \,.$$

To generate this solution first we apply the above method in Case 1 with λ_2. If it works, and it yields a second solution, then it is y_2 of the previous form with $d_{-1} = 0$. Otherwise we apply the method in Case 2 and take

$$y_2(x) = \left. \frac{\partial}{\partial \lambda} [(\lambda - \lambda_2) y(\lambda, x)] \right|_{\lambda = \lambda_2} .$$

5.18 Problems

1. Find a solution of the differential equation
$$x^2y'' - xy' + y = 0$$
near to $x_0 = 0$.
2. Find the general solution of the differential equation
$$3x^2y'' - xy' + y = 0$$
near to $x_0 = 0$.
3. Find the general solution of the differential equation
$$(1 + x^2)y'' - 4xy' + 6y = 0$$
near to $x_0 = 0$.
4. Find a solution of the differential equation
$$x^2y'' + (x^2 - 2x)y' + 2y = 0$$
near to $x_0 = 0$.

5.19 The Laplace transform

Let $f : [0, +\infty[\to \mathbb{R}$ be a function. We define
$$\mathcal{L}(f)(s) = \int_0^{+\infty} e^{-sx} f(x)\, dx \tag{5.58}$$
for those values of s, for which the improper integral exists. The function $\mathcal{L}(f)$ is called the *Laplace transform* of f. In the following theorem we summarize the most important properties of the Laplace transform. The statements of this theorem can be verified by direct calculation using the properties of improper integrals.

Theorem 5.19.1. *Let $f, g : [0, +\infty[\to \mathbb{R}$ be continuous functions. Then for every constants $\omega > 0, \lambda, \mu, a$ and positive integer n we have the following statements:*

1. *If the Laplace transforms of f and g exist, then so does the Laplace transform of $\lambda f + \mu g$, and*
$$\mathcal{L}(\lambda f + \mu g)(s) = \lambda\mathcal{L}(f)(s) + \mu\mathcal{L}(g)(s)\,,$$
2. *If the Laplace transform of f exists, then so does the Laplace transform of $e^{ax} f(x)$, and*
$$\mathcal{L}\big(e^{ax} f(x)\big)(s) = \mathcal{L}(f)(s - a)\,,$$

3. *If the Laplace transform of f is n-times differentiable, then the Laplace transform of $x^n f(x)$ exists, and*

$$\big(x^n f(x)\big)(s) = (-1)^n \frac{d^n}{ds^n}\mathcal{L}(f)(s),$$

4. *If the limit $\lim_{x\to 0+0} \frac{f(x)}{x}$ exists, and the Laplace transform of f is integrable on $[0, +\infty[$, then the Laplace transform of $\frac{1}{x}f(x)$ exists, then*

$$\mathcal{L}\Big(\frac{1}{x}f(x)\Big)(s) = \int_s^{+\infty} \mathcal{L}(f)(t)\,dt,$$

5. *If the Laplace transform of f exists, then so does the Laplace transform of $\int_0^x f(t)dt$ for each x, and*

$$\mathcal{L}\Big(\int_0^x f(t)\,dt\Big)(s) = \frac{1}{s}\mathcal{L}(f)(s),$$

6. *If f is periodic with period ω, and its Laplace transform exists, then*

$$\mathcal{L}(f)(s) = \frac{\int_0^\omega e^{-sx} f(x)\,dx}{1 - e^{-\omega s}}.$$

From the point of view of differential equations the following theorem is of basic importance.

Theorem 5.19.2. *Let n be a positive integer, and let $f : [0, +\infty[\to \mathbb{R}$ be an n-times continuously differentiable function which satisfies*

$$\lim_{x\to+\infty} f^{(k)}(x)e^{-sx} = 0 \tag{5.59}$$

for each s in the domain of $\mathcal{L}(f)$, and for every $k = 0, 1, \ldots, n-1$. Then we have

$$\mathcal{L}(f^{(n)})(s) = s^n\mathcal{L}(f)(s) - \sum_{k=0}^{n-1} s^{n-1-k} f^{(k)}(0). \tag{5.60}$$

Proof. The statement can be proved applying repeated integration by parts. □

We have the special cases

$$\mathcal{L}(f') = s\mathcal{L}(f)(s) - f(0)$$
$$\mathcal{L}(f'') = s^2\mathcal{L}(f)(s) - sf(0) - f'(0).$$

Using this theorem n-th order linear differential equations and initial value problems can be transformed into algebraic equations. We consider the following example:

$$y' - 5y = e^{5x}, \quad y(0) = 0. \tag{5.61}$$

Supposing that y satisfies the conditions above we have for the Laplace transform of y:

$$\mathcal{L}(y'') - 5\mathcal{L}(y) = \mathcal{L}(e^{5x}).$$

We get

$$\mathcal{L}(e^{5x}) = \int_0^{+\infty} e^{-sx}\, e^{5x}\, dx = \int_0^{+\infty} e^{(5-s)x}\, dx = \frac{1}{s-5},$$

hence, by the above theorem and by the initial condition, it follows

$$s\mathcal{L}(f)(s) - 5\mathcal{L}(f)(s) = \frac{1}{s-5},$$

that is

$$\mathcal{L}(f)(s) = \frac{1}{(s-5)^2}. \tag{5.62}$$

Consequently, we may find a solution of the given initial value problem, if we find a function, whose Laplace transform is $\frac{1}{(s-5)^2}$. This leads to the problem of the inverse Laplace transform. If F is a function defined on some subset of the reals, then the function $f : [0, +\infty[\to \mathbb{R}$ is called its *inverse Laplace transform*, denoted by $\mathcal{L}^{-1}(F)$, if $\mathcal{L}(f) = F$. Unfortunately, there is no general rule for the computation of the inverse Laplace transform, however, this computation is possible in some special classes of functions. One obvious way is to have a table of the Laplace transforms of the basic functions, and then this table can be used in the "reverse direction" for determining inverse Laplace transforms. In fact, there are tables of Laplace transforms, one of them is in the end of this section. Another way is to specify some function classes, for which we can compute the inverse Laplace transforms using algebraic tools. We exhibit some simple methods.

First we show, that for each natural number k and real number a we have

$$\mathcal{L}(x^k e^{ax}) = (-1)^{k+1}\frac{k!}{(a-s)^{k+1}}, \tag{5.63}$$

which can be checked easily, using integration by parts. Applying this formula we determine the inverse Laplace transform of the function

$$F(s) = \frac{s+3}{(s-2)(s+1)}.$$

First we decompose F as a sum of partial fractions:

$$\frac{s+3}{(s-2)(s+1)} = \frac{A}{s-2} + \frac{B}{s+1}.$$

We obtain $A = \frac{5}{3}$, $B = \frac{-2}{3}$, hence

$$\mathcal{L}^{-1}(F)(x) = \frac{5}{3}\mathcal{L}^{-1}\Big(\frac{1}{s-2}\Big) - \frac{2}{3}\mathcal{L}^{-1}\Big(\frac{1}{s+1}\Big) = \frac{5}{3}e^{2x} - \frac{2}{3}e^{-x},$$

by equation (5.63). We have seen in Section 5.13 that every exponential polynomial is a linear combination of exponential monomials of the form $x^k e^{ax}$, hence, by formula (5.63), we can find the inverse Laplace transform of every exponential polynomial.

We can use this observation to solve the initial value problem (5.61) above. Namely, from (5.62) we have, using (5.63):

$$y(x) = \mathcal{L}^{-1}\Big(\frac{1}{(s-5)^2}\Big) = xe^{5x},$$

which is the solution of the problem (5.61).

The concept of Laplace transform and inverse Laplace transform can be extended to complex valued functions. Indeed, if $f;[0,+\infty[\to \mathbb{C}$ is given, then its Laplace transform can be defined as

$$\mathcal{L}(f) = \mathcal{L}(\operatorname{Re} f) + i\mathcal{L}(\operatorname{Im} f),$$

whenever the Laplace transform of its real part $\operatorname{Re} f$ and imaginary part $\operatorname{Im} f$ exists. The inverse Laplace transform of a complex valued function is meant in the same way. Using this extension we can deal with the Laplace transform of trigonometric functions easily, as they turn to be exponential polynomials. Indeed, we have

$$\cos ax = \frac{e^{iax} + e^{-iax}}{2}, \quad \sin ax = \frac{e^{iax} - e^{-iax}}{2i}.$$

It follows

$$\mathcal{L}(\cos ax) = \frac{s}{s^2+a^2}, \quad \mathcal{L}(\sin ax) = \frac{1}{s^2+a^2}. \tag{5.64}$$

These formulas can be utilized when computing inverse Laplace transform using partial fraction decomposition.

5.20 Problems

1. Find the Laplace transform of each of the following functions:
   i) $f(x) = 6e^{-5x} + e^{3x} + 5x^3 - 9$
   ii) $f(x) = 4\cos 4x - 9\sin 4x + 2\cos 10x$
   iii) $f(x) = 3\sinh 2x + 3\sin 2x$

iv) $f(x) = e^{3x} + \cos 6x - e^{3x} \cos 6x$

2. Find the Laplace transform of each of the following functions:

   i) $f(x) = x \cosh 3x$
   ii) $f(x) = x^2 \sin 2x$
   iii) $f(x) = x^{\frac{3}{2}}$
   iv) $f(x) = (10x)^{\frac{3}{2}}$

3. Find the inverse Laplace transform of each of the following functions:

   i) $F(s) = \frac{6}{s} - \frac{1}{s-8} + \frac{4}{s-3}$
   ii) $F(s) = \frac{19}{s+2} - \frac{1}{3s-5} + \frac{7}{s^2}$
   iii) $F(s) = \frac{6s}{s^2+25} + \frac{3}{s^2+25}$
   iv) $F(s) = \frac{8}{3s^2+12} + \frac{3}{s^2-49}$

4. Find the inverse Laplace transform of each of the following functions:

   i) $F(s) = \frac{6s-5}{s^2+7}$
   ii) $F(s) = \frac{1-3s}{s^2+8s+2}$
   iii) $F(s) = \frac{3s-2}{2s^2-6s-2}$
   iv) $F(s) = \frac{s+7}{s^2-3s-10}$

5. Find the inverse Laplace transform of each of the following functions:

   i) $F(s) = \frac{86s-78}{(s+3)(s-4)(5s-1)}$
   ii) $F(s) = \frac{2-5s}{(s-6)(s^2+11)}$
   iii) $F(s) = \frac{25}{s^3(2s^2+4s+5)}$

6. Solve the following initial value problems using Laplace transform:

   i) $y'' - 10y' + 9y = 5x,\ \ y(0) = -1,\ \ y'(0) = 2$
   ii) $2y'' + 3y' - 2y = xe^{-2x},\ \ y(0) = 0,\ \ y'(0) = -2$
   iii) $y'' - 6y' + 15y = 2 \sin 3x,\ \ y(0) = -1,\ \ y'(0) = -4$

5.21 The Fourier transform of exponential polynomials

Besides Laplace transform another fundamental transform which can be utilized to solve differential equations is the *Fourier transform.* In the classical Fourier analysis the Fourier transform is defined for functions satisfying different integrability conditions. Unfortunately, in most cases the solutions of important differential equations, like exponential polynomials, do not satisfy such conditions. It turns out, however, that a special "Fourier-like transform" can be introduced for exponential polynomials, and this transform has similar convenient properties to those of the Fourier transform. In this section we introduce this transform and show how to use it in solving differential equations. Later we shall see that this transform can be used also in the case of partial differential equations, hence we give here the definition for functions of several variables.

The definition for exponential polynomials given in Section 5.13 can be generalized as follows. Let n be a positive integer. The function $m : \mathbb{R}^n \to \mathbb{C}$ is called an *exponential*, if it has the form

$$m(x) = \exp\langle \lambda, x \rangle \tag{5.65}$$

for each x in $\mathbb{R}^n$, where λ is in $\mathbb{C}^n$. Hence, if $x = (x_1, x_2, \ldots, x_n)$ and $\lambda = (\lambda_1, \lambda_2, \ldots, \lambda_n)$, then we have

$$m(x) = \exp \sum_{j=1}^{n} \lambda_j x_j \,.$$

The function $P : \mathbb{R}^n \to \mathbb{C}$ is called a *polynomial*, if it has the form

$$P(x) = \sum_{|\alpha| \leqslant N} c_\alpha x^\alpha \tag{5.66}$$

for each x in $\mathbb{R}^n$, where N is a natural number, and c_α is a complex number for each α multi-index, with $|\alpha| \leqslant N$. We recall that a *multi-index* is an element $\alpha = (\alpha_1, \alpha_2, \ldots, \alpha_n)$ of $\mathbb{N}^n$, and $|\alpha| = \alpha_1 + \alpha_2 + \cdots + \alpha_n$. Addition and inequalities between multi-indices is defined componentwise, further

$$x^\alpha = x_1^{\alpha_1} x_2^{\alpha_2} \ldots x_n^{\alpha_n}$$

for each x in $\mathbb{R}^n$ and multi-index α in $\mathbb{N}^n$.

The function $\varphi : \mathbb{R}^n \to \mathbb{C}$ is called an *exponential monomial*, if it has the form

$$\varphi(x) = P(x)m(x) \tag{5.67}$$

for each x in $\mathbb{R}^n$, where $P : \mathbb{R}^n \to \mathbb{C}$ is a polynomial and $m : \mathbb{R}^n \to \mathbb{C}$ is an exponential. Linear combinations of exponential monomials are called *exponential polynomials*. We have the following important result.

Theorem 5.21.1. *Every non-zero exponential polynomial φ can be written uniquely in the following form:*

$$\varphi(x) = \sum_{j=1}^{k} P_k(x) m_k(x) \tag{5.68}$$

for each x in $\mathbb{R}^n$, where the exponentials $m_j : \mathbb{R}^n \to \mathbb{C}$ are different, and the polynomials $P_j : \mathbb{R}^n \to \mathbb{C}$ are non-zero ($j = 1, 2, \ldots, k$).

The representation (5.68) is called the *canonical form* of the exponential polynomial φ. This theorem makes it possible to introduce the new transform for exponential polynomials. Namely, by the above theorem, it follows that given an exponential polynomial a unique polynomial-valued function can be defined on the set of all exponentials such that the value of this function at each exponential is equal to the polynomial coefficient of this exponential in the unique representation of the given exponential polynomial of the above form. To make this more formal let $\mathcal{P}(\mathbb{R}^n)$ denote the set of all complex polynomials on $\mathbb{R}^n$ and $\mathcal{EP}(\mathbb{R}^n)$ the set of all exponential polynomials on $\mathbb{R}^n$. For each φ in $\mathcal{EP}$ and for each complex vector λ in $\mathbb{C}^n$ let $\widehat{\varphi}(\lambda)$ denote the unique polynomial in $\mathcal{P}(\mathbb{R}^n)$ such that

$$\varphi(x) = \sum_{\lambda \in \mathbb{C}^n} \widehat{\varphi}(\lambda)(x) \exp\langle \lambda, x \rangle \tag{5.69}$$

holds for each x in $\mathbb{R}^n$. We note that the sum is obviously finite for every exponential polynomial φ. In the following theorem we list the most important properties of the transform $\varphi \mapsto \widehat{\varphi}$ which will be called the *Fourier transformation*, while $\widehat{\varphi}$ is called the *Fourier transform* of the exponential polynomial φ. For each x, y in $\mathbb{R}^n$ and for every function $f : \mathbb{R}^n \to \mathbb{C}$ we define the *translate* of f by y, respectively the *inversion* of f as

$$\tau_y f(x) = f(x+y), \text{ respectively } \check{f}(x) = f(-x)\,.$$

Theorem 5.21.2. *We have the following statements:*

i) The Fourier transformation $\varphi \mapsto \widehat{\varphi}$ is a linear mapping from $\mathcal{EP}$ into the set of all finitely supported polynomial-valued functions on $\mathbb{C}^n$.
ii) $\widehat{p}(\lambda) = 0$, whenever $\lambda \neq 0$, and $\widehat{p}(0) = p$.
iii) $(pf)\widehat{\ }(\lambda) = p \cdot \widehat{f}(\lambda)$.
iv) $(\tau_y f)\widehat{\ }(m) = m(y) \cdot (\tau_y \widehat{f})(m)$.
v) $(\check{f})\widehat{\ }(m) = [\widehat{f}(\check{m})]\check{\ }$.

Obviously, by Theorem 5.21.1, the following “Inversion Theorem” holds:

Theorem 5.21.3. *(Inversion Theorem) For every exponential polynomial $\varphi : \mathbb{R}^n \to \mathbb{C}$ we have*

$$\varphi(x) = \sum_{\lambda in \mathbb{C}} \widehat{\varphi}(\lambda)(x) e^{\langle \lambda, x \rangle} \tag{5.70}$$

for each x in $\mathbb{R}$.

The proof of these statements is an easy calculation and it is left to the reader. We note that the fourth property, called *translation covariance*, is very important from the point of view of differential equations, as we shall see it in our next theorem. In the theorem we use the following notation: for each $j = 1, 2 \ldots, n$ the symbol ∂_j denotes the partial differential operator with respect to the j-th variable defined on the set of all complex valued differentiable functions on $\mathbb{R}^n$. Using the notation $\partial = (\partial_1, \partial_2, \ldots, \partial_n)$ for each multi-index α we have

$$\partial^\alpha = \partial_1^{\alpha_1} \partial_2^{\alpha_2} \ldots \partial_n^{\alpha_n},$$

which acts on $|\alpha|$-times differentiable functions in the obvious way. If P is a polynomial of the form (5.66), then we define

$$P(\partial) = \sum_{|\alpha| \leqslant N} c_\alpha \partial^\alpha . \tag{5.71}$$

Then $P(\partial)$ is a *linear differential operator with constant coefficients* which acts on N-times differentiable functions in the obvious way. We note that the constant term of P is α_0, where $0 = (0, 0, \ldots, 0)$ is the zero multi-index, and the corresponding term in $P(\partial)$ is $\alpha_0 \cdot \partial^0$ which is meant as $\alpha_0 \cdot I$, where I denotes the identity operator: $If = f$ for every function f.

Theorem 5.21.4. *Let P be a complex polynomial on $\mathbb{R}^n$, and $\varphi : \mathbb{R}^n \to \mathbb{C}$ is an exponential polynomial. Then for any λ in $\mathbb{C}^n$ we have*

$$(P(\partial)f)\hat{}(\lambda) = P(\partial + \lambda \cdot I)\widehat{f}(\lambda) .$$

The theorem can be verified by direct calculation. In the case $n = 1$ we have the following corollary. Here $D = \partial_1$ denotes the ordinary differential operator.

Theorem 5.21.5. *Let $f : \mathbb{R} \to \mathbb{C}$ be an exponential polynomial and k a non-negative integer. Then for any complex λ we have*

$$(D^k f)\hat{}(\lambda) = (D + \lambda \cdot I)^k \widehat{f}(\lambda) .$$

As an application we show how to find all solutions of inhomogeneous linear differential equations of the form

$$P(D)y = \varphi \tag{5.72}$$

where P is a complex polynomial of degree N on $\mathbb{R}$ and φ is a given exponential polynomial. We recall that here P is the characteristic polynomial of the given equation as it was defined in Section 5.7. It turns out that (3.1) always has an exponential polynomial solution. As we have seen in Section 5.7 all solutions of the corresponding homogeneous equation are exponential polynomials, it follows that all solutions of (3.1) are exponential polynomials. Let y denote any solution of (3.1). Applying Fourier transformation on both sides of (3.1) we have that

$$P(D+\lambda)\widehat{y}(\lambda)=\widehat{\varphi}(\lambda)$$

holds for each complex λ. Here $\widehat{y}(\lambda)$ and $\widehat{\varphi}(\lambda)$ are polynomials, and $\widehat{\varphi}(\lambda)=0$ for all but finitely many values of λ. Hence, the problem of solving (3.1) is reduced to the problem of finding all polynomial solutions q of an equation of the form

$$P(D+\lambda\cdot I)q=p\,, \tag{5.73}$$

where p is a given polynomial.

First we suppose that

$$p(x)=\sum_{k=0}^{d}c_kx^k\,,$$

where d is a natural number and $c_d\neq 0$. Then (5.73) is equivalent to the equation

$$\sum_{j=0}^{N}\frac{P^{(j)}(\lambda)}{j!}D^jq(x)=\sum_{k=0}^{d}c_kx^k\,. \tag{5.74}$$

Suppose that λ is a zero of order l of P, where l is a natural number with $0\leqslant l\leqslant N$, that is, $P^{(j)}(\lambda)=0$ for $j=0,1,\ldots,l-1$, and $P^{(l)}(\lambda)\neq 0$. Obviously, in the case $l=0$ we simply mean that $P(\lambda)\neq 0$. Now we put $r=D^lq$ and we obtain from (5.74)

$$\sum_{j=0}^{N-l}\frac{P^{(j+l)}(\lambda)}{(j+l)!}D^jr(x)=\sum_{k=0}^{d}c_kx^k\,. \tag{5.75}$$

It follows that $\deg r\leqslant d$. We let $r(x)=\sum_{k=0}^{d}b_kx^k$, then substitution into (5.75) yields

$$\sum_{j=0}^{N-l}\sum_{i=j}^{d}\frac{P^{(j+l)}(\lambda)}{(j+l)!}j!\binom{i}{j}b_ix^{i-j}=\sum_{k=0}^{d}c_kx^k\,. \tag{5.76}$$

Comparing the coefficients of x^i we have

$$\sum_{j=0}^{\min(N-l,d-i)}\frac{P^{(j+l)}(\lambda)}{(j+l)!}j!\binom{i+j}{j}b_{i+j}=c_i\qquad (i=0,1,\ldots,d)\,. \tag{5.77}$$

This system of linear equations for the b_i's has triangular form, hence it is easily solvable. If $N - l > d$, then the system has the following form:

$$
\begin{aligned}
c_d &= \frac{P^{(l)}(\lambda)}{l!}\, b_d \\
c_{d-1} &= \frac{P^{(l)}(\lambda)}{l!}\, b_{d-1} + \frac{P^{(l+1)}(\lambda)}{(l+1)!}\, d \cdot b_d \\
&\ \ \vdots \\
c_0 &= \frac{P^{(l)}(\lambda)}{l!}\, b_0 + \frac{P^{(l+1)}(\lambda)}{(l+1)!}\, b_1 + \frac{P^{(l+2)}(\lambda)}{(l+2)!}\, 2!\, b_2 + \cdots + \frac{P^{(l+d)}(\lambda)}{(l+d)!}\, d!\, b_d\,,
\end{aligned}
$$

which is obviously uniquely solvable. If $N - l \leqslant d$, then we have the following form from (5.77):

$$
\begin{aligned}
c_d &= \frac{P^{(l)}(\lambda)}{l!}\, b_d \\
c_{d-1} &= \frac{P^{(l)}(\lambda)}{l!}\, b_{d-1} + \frac{P^{(l+1)}(\lambda)}{(l+1)!}\, d \cdot b_d \\
&\ \ \vdots \\
c_{d-(N-l)} &= \frac{P^{(l)}(\lambda)}{l!}\, b_{d-(N-l)} + \frac{P^{(l+1)}(\lambda)}{(l+1)!} \binom{d-(N-l)+1}{1} b_{d-(N-l)+1} \\
&\quad + \cdots + \frac{P^{(N)}(\lambda)}{N!}\, (N-l)! \binom{d}{N-l} b_d \\
c_{d-(N-l)-1} &= \frac{P^{(l)}(\lambda)}{l!}\, b_{d-(N-l)-1} + \frac{P^{(l+1)}(\lambda)}{(l+1)!} \binom{d-(N-l)}{1} b_{d-(N-l)} \\
&\quad + \cdots + \frac{P^{(N)}(\lambda)}{N!}\, (N-l)! \binom{d-1}{N-l} b_{d-1} \\
&\ \ \vdots
\end{aligned}
$$

$$c_0 = \frac{P^{(l)}(\lambda)}{l!}\, b_0 + \frac{P^{(l+1)}(\lambda)}{(l+1)!}\, b_1 + \cdots + \frac{P^{(N)}(\lambda)}{N!}\, (N-l)!\, b_{N-l}\,,$$

which has a unique solution, too.

In the case $p = 0$ we see from (5.74) that $q \neq 0$ implies $P(\lambda) = 0$, hence λ is a characteristic root with multiplicity $l \geqslant 1$. Then, by (5.75), it follows $r = 0$, hence q is an arbitrary polynomial of degree at most $l - 1$.

It is obvious that the exponential polynomial y for which $\widehat{y} = q$, where q is the polynomial whose coefficients are determined by the system of equations (5.77) is a solution of (5.72). Hence we have proved the following theorem.

Theorem 5.21.6. *Let $\varphi : \mathbb{R} \to \mathbb{C}$ be an exponential polynomial. Then all solutions of the differential equation* (5.72) *are exponential polynomials. Further, if y is a solution, and the complex number λ is a characteristic root of order l of* (5.72) *(l is a natural number), then the coefficients b_j of the polynomial $D^k\widehat{y}(\lambda)$ satisfy the system of equations* (5.77), *where the c_i's are the coefficients of $\widehat{\varphi}(\lambda)$.*

Summarizing the results we can solve (5.72) as follows. First we determine the zeros of P together with their multiplicities. Suppose that the support of $\widehat{\varphi}(\lambda)$ is the finite set $\lambda_1, \lambda_2, \ldots, \lambda_s$. If $1 \leqslant i \leqslant s$, then we determine the value l such that $P^{(j)}(\lambda_i) = 0$ for $j = 0, 1, \ldots, l-1$, and $P^{(l)}(\lambda_i) \neq 0$, and we solve the system of equations (5.77) with $\lambda = \lambda_i$. The constants c_j are the coefficients of $\widehat{\varphi}(\lambda_i)$. From the solution we obtain the polynomial $q_i = \widehat{y}(\lambda_i)$. We do the same for $i = 1, 2, \ldots, s$. Finally, the solution is

$$y(x) = \sum_{i=1}^{s} q_i(x) e^{\lambda_i x} + \sum_{j=1}^{t} p_j(x) e^{\mu_j x}\,, \tag{5.78}$$

where $\mu_1, \mu_2, \ldots, \mu_t$ are the characteristic roots of the polynomial P with multiplicities $n_1, n_2, \ldots, n_t$ ($n_1 + n_2 + \cdots + n_t = N$), and p_j is an arbitrary polynomial of degree at most $n_j - 1$ for $j = 0, 1, \ldots, t$.

As an illustration we solve the differential equation

$$y'' - y = x^2 e^x - x\cos x + 1\,. \tag{5.79}$$

The characteristic polynomial is $P(\lambda) = \lambda^2 - 1$, hence the characteristic roots are $\mu_1 = 1$ and $\mu_2 = -1$. Applying Fourier transformation on both sides we obtain

$$q'' + 2\lambda q' + (\lambda^2 - 1)q = \begin{cases} x^2 & \text{if } \lambda = 1 \\ -\frac{x}{2} & \text{if } \lambda = i \\ -\frac{x}{2} & \text{if } \lambda = -i \\ 1 & \text{if } \lambda = 0 \\ 0 & \text{otherwise.} \end{cases}$$

Here $q = \widehat{y}(\lambda)$ and we used the canonical form of the exponential polynomial on the right hand side of equation (5.79):

$$x^2 e^x - x\cos x + 1 = x^2 e^x - \frac{x}{2}e^{ix} - \frac{x}{2}e^{-ix} + 1\,.$$

For $\lambda = 1$ we have from (5.77)

$$\begin{aligned} 1 &= 2b_2 \\ 0 &= 2b_1 + 2b_2 \\ 0 &= 2b_0 + b_1\,, \end{aligned}$$

which implies $b_2 = \frac{1}{4}, b_1 = -\frac{1}{2}, b_0 = \frac{1}{4}$. Hence

$$\widehat{y}(1)(x) = \frac{1}{6}x^3 - \frac{1}{4}x^4 + \frac{1}{4}x + c.$$

For $\lambda = i$ the system (5.77) implies

$$\begin{aligned} -\frac{1}{2} &= -2b_1 \\ 0 &= -2b_0 + 2ib_1\,, \end{aligned}$$

which gives

$$\widehat{y}(i)(x) = \frac{1}{4}x + \frac{i}{4}.$$

In the case $\lambda = -i$ we get similarly

$$\widehat{y}(-i)(x) = \frac{1}{4}x - \frac{i}{4}.$$

For $\lambda = 0$ we have $\widehat{y}(0)(x) = -1$, and finally, if $\lambda \neq 1, \lambda \neq \pm i, \lambda \neq 0$, then $\widehat{y}(\lambda) \neq 0$ implies $\lambda = -1$, and $\widehat{y}(-1)(x) = d$, a constant. Thus the general solution of the differential equation (5.79) is

$$y(x) = ce^x + de^{-x} + (\frac{1}{6}x^3 - \frac{1}{4}x^4 + \frac{1}{4}x)e^x + \frac{1}{2}x\cos x - \frac{1}{2}\sin x - 1\,,$$

where c, d are arbitrary constants.

The second property in Theorem 5.21.2 makes it possible to handle linear differential equations with polynomial coefficients, too. As an example we determine all exponential polynomial solutions of the differential equation

$$(x^2-1)y''-(3x+1)y'-(x^2-x)y=0\,. \tag{5.80}$$

Applying Fourier transformation we obtain

$$(x^2-1)q''+(2\lambda x^2-3x-2\lambda-1)q'+\big(x^2(\lambda^2-1)-x(3\lambda-1)-\lambda^2-\lambda\big)q=0\,,$$

where $q=\widehat{y}(\lambda)$. From this equation, by comparing the leading coefficients of x, we infer that $q\neq 0$ implies $\lambda=\pm 1$. If $\lambda=1$, then we have

$$(x^2-1)q''+(2x^2-3x-3)q'-(2x+2)q=0\,.$$

Here $q\neq 0$ implies $\deg q\leqslant 1$, but substitution gives that there is no non-zero solution with this property. If $\lambda=-1$, then we have

$$(x^2-1)q''+(-2x^2-3x+1)q'+4xq=0\,.$$

It follows that $q\neq 0$ implies $\deg q\leqslant 2$. Substituting $q(x)=x^2+ax+b$ into the equation we get $a=2, b=1$, and $q(x)=(x+1)^2$ is a solution. Hence the exponential polynomial

$$y(x)=(x+1)^2e^{-x}$$

is a solution of the differential equation (5.80). Using the reduction method presented in Section 5.9 we can reduce the problem to a first order linear differential equation which can be solved.

5.22 Problems

1. Solve the following differential equations.
   i) $y'''-3y''+3y'-y=x\sin x$
   ii) $y''+4y=x\sin 2x$
   iii) $xy'''-y''-xy'+y=8x^2e^x$
   iv) $x^4y^{(iv)}+6x^3y'''+2x^2y''-4xy'+4y=12x^2$
   v) $x^3y'''-x^2(x+3)y''+2x(x+3)y'-2(x+3)y=-4x^4$
   vi) $x^3y'''-3x^2y''+6xy'-6y=2x$
2. Solve the following differential equations.
   i) $x^2y''-xy'+y=0$
   ii) $xy'''-(x-3)y''-(x-2)y'+(x-1)y=-4e^{-x}$
   iii) $xy'''+x^2y''-2xy'+2y=x^2$
   iv) $xy'''-y''-xy'+y=x^2$

Chapter 6

FIRST ORDER PARTIAL DIFFERENTIAL EQUATIONS

6.1 Homogeneous linear partial differential equations

Throughout this chapter n denotes a positive integer. Let $\Omega \subseteq \mathbb{R}^n$ be a non-empty open set, and let $f : \Omega \to \mathbb{R}^n$ be a continuous function which is non-identically zero. The

$$\langle f(x), u'(x) \rangle = 0 \tag{6.1}$$

equation is called *first order homogeneous linear partial differential equation*. With the notation $f = (f_1, f_2, \dots, f_n)$ (6.1) means the equation

$$\sum_{k=1}^{n} f_k(x) \partial_k u(x) = 0 \, .$$

We say that u is *a solution of* (6.1) *on* Ω_0, if Ω_0 is a non-empty open subset of Ω, $u : \Omega_0 \to \mathbb{R}$ is a continuously differentiable function, and for each x in Ω_0 we have (6.1). By the Chain Rule, it follows immediately, that if k is a positive integer and $u_1, u_2, \dots, u_k$ are solutions of (6.1) on some open set $\Omega_0 \subseteq \Omega$, further u is functionally expressible by these functions on Ω_0, then u is also a solution of (6.1) on Ω_0.

Theorem 6.1.1. *If the function f in* (6.1) *does not vanish at any point in the non-empty open set $\Omega_0 \subseteq \Omega$, then any n solutions of* (6.1) *on Ω_0 are functionally dependent.*

Proof. Let $\varphi_1, \varphi_2, \dots, \varphi_n$ be solutions of (6.1) on the set Ω_0. If x is in Ω_0, then we have

$$\sum_{k=1}^{n} f_k(x) \partial_k \varphi_i(x) = 0$$

for $i = 1, 2, \dots, n$ which is a homogeneous linear system of equations for the unknowns $f_1(x), f_2(x), \dots, f_n(x)$, where $f = (f_1, f_2, \dots, f_n)$. As f does not vanish on Ω_0, this system has a non-trivial solution, hence the determinant of its fundamental matrix is zero:

$$\det\big(\partial_k \varphi_i(x)\big) = 0\,.$$

As x is arbitrary, this means that $\varphi_1, \varphi_2, \ldots, \varphi_n$ are functionally dependent on Ω_0. □

By the theorem, under the given conditions equation (6.1) can have at most $n-1$ functionally independent solutions. In fact, if it has $n-1$ functionally independent solutions, then these solutions play the role of a kind of "basis" in the space of all solutions in the sense as it is expressed in the following theorem.

Theorem 6.1.2. *If the function f in* (6.1) *does not vanish at any point in the non-empty open set $\Omega_0 \subseteq \Omega$, further $\{\varphi_1, \varphi_2, \ldots, \varphi_{n-1}\}$ is a fundamental system of* (6.1) *on Ω_0, then each solution of* (6.1) *on Ω_0 is functionally expressible by the functions $\varphi_1, \varphi_2, \ldots, \varphi_{n-1}$.*

Proof. As the functions $\varphi_1, \varphi_2, \ldots, \varphi_{n-1}$ are independent on Ω_0, hence it is enough to show that if the function u is a solution of (6.1) on Ω_0, then $\varphi_1, \varphi_2, \ldots, \varphi_{n-1}, u$ are functionally dependent on Ω_0. However, this follows from the previous theorem. □

Using the notation of the above theorems we say that $\{\varphi_1, \varphi_2, \ldots, \varphi_{n-1}\}$ is a *fundamental system* of (6.1) on Ω_0, if $\varphi_1, \varphi_2, \ldots, \varphi_{n-1}$ are functionally independent solutions of (6.1) on Ω_0.

Theorem 6.1.3. *Let $f = (f_1, f_2, \ldots, f_n)$ be given in* (6.1), *and let Ω_0 be a non-empty open subset of Ω such that f_1 does not vanish at any point of it. The continuously differentiable function $\varphi : \Omega_0 \to \mathbb{R}$ is a solution of* (6.1) *on Ω_0 if and only if φ is a first integral on Ω_0 of the system of differential equations*

$$y_i'(t) = \frac{f_{i+1}(t, y_1, y_2, \ldots, y_{n-1})}{f_1(t, y_1, y_2, \ldots, y_{n-1})}\,, \qquad (i = 1, 2, \ldots, n-1)\,. \tag{6.2}$$

Proof. Suppose that φ is a solution of (6.1) on Ω_0, further let $y : I \to \mathbb{R}^{n-1}$ be a solution of (6.2) such that for each t in I the point $\big(t, y(t)\big)$ belongs to Ω_0. Let $y = (y_1, y_2, \ldots, y_{n-1})$, then we have

$$\frac{d}{dt}\varphi\big(t, y_1(t), \ldots, y_{n-1}(t)\big)$$

$$= \partial_1\varphi\big(t, y_1(t), \dots, y_{n-1}(t)\big) + \sum_{i=1}^{n-1} y_i'(t)\partial_{i+1}\varphi\big(t, y_1(t), \dots, y_{n-1}(t)\big)$$

$$= \partial_1\varphi\big(t, y_1(t), \dots, y_{n-1}(t)\big)$$

$$+ \sum_{i=1}^{n-1} \frac{f_{i+1}\big(t, y_1(t), \dots, y_{n-1}(t)\big) \cdot \partial_{i+1}\varphi\big(t, y_1(t), \dots, y_{n-1}(t)\big)}{f_1\big(t, y_1(t), \dots, y_{n-1}(t)\big)}$$

$$= \sum_{i=1}^{n} \frac{f_i\big(t, y_1(t), \dots, y_{n-1}(t)\big) \cdot \partial_i\varphi\big(t, y_1(t), \dots, y_{n-1}(t)\big)}{f_1\big(t, y_1(t), \dots, y_{n-1}(t)\big)} = 0\,.$$

Conversely, let φ be a first integral of (6.1) on Ω_0, and let $(x_1, x_2, \dots, x_n)$ be an arbitrary point in Ω_0. By Theorem 1.5.1 of Peano, there exists a solution $y : I \to \mathbb{R}^{n-1}$ of (6.1) such that with the notation $y = (y_1, y_2, \dots, y_{n-1})$ we have $y_i(x_1) = x_{i+1}$ for $i = 1, 2, \dots, n-1$. Here x_1 is an arbitrary point in I. Then we obtain

$$\sum_{i=1}^{n} f_i(x_1, \dots, x_n)\, \partial_i\varphi(x_1, \dots, x_n)$$

$$= f_1\big(x_1, y_1(x_1), \dots, y_{n-1}(x_1)\big)$$

$$\cdot \sum_{i=1}^{n} \frac{f_i\big(x_1, y_1(x_1), \dots, y_{n-1}(x_1)\big)}{f_1\big(x_1, y_1(x_1), \dots, y_{n-1}(x_1)\big)}\, \partial_i\varphi\big(x_1, y_1(x_1), \dots, y_{n-1}(x_1)\big)$$

$$= f_1\big(x_1, y_1(x_1), \dots, y_{n-1}(x_1)\big) \cdot \frac{d}{dt}\varphi\big(t, y_1(t), \dots, y_{n-1}(t)\big)\big|_{t=x_1} = 0\,,$$

which proves our theorem. □

The system (6.2) of ordinary differential equations is called the *characteristic system of equations* of the first order homogeneous partial differential equation (6.1).

Theorem 6.1.4. *If the function f in (6.1) is continuously differentiable and it does not vanish at any point of the non-empty open set $\Omega_0 \subseteq \Omega$, then each point of Ω_0 has a neighborhood in which there exists a fundamental system of (6.1).*

Proof. Let x_0 be a point in Ω_0. As $f(x_0) \neq 0$, we may suppose that $f_1(x_0) \neq 0$, where $f = (f_1, f_2, \dots, f_n)$. Then x_0 has a neighborhood $U \subseteq \Omega_0$ such that $f_1(x) \neq 0$, whenever x is in U. By the previous theorem, it is enough to show that in some neighborhood of x_0 the characteristic system of equations (6.2) has $n-1$ independent first integrals. But this is a consequence of Theorem 4.3.1. □

By our above considerations, we can obtain each solution of the first order homogeneous linear partial differential equation in some neighborhood of any point in the following way: first we determine $n-1$ functionally independent first integrals of the characteristic system of equations in some neighborhood of the given point, then – roughly speaking – we substitute these functions in an arbitrary continuously differentiable function of $n-1$ variables. This is usually called the *general solution* of the first order homogeneous linear partial differential equation.

As an illustration we solve the partial differential equation

$$x\,\frac{\partial u}{\partial x} + \frac{x^2(x+2y)}{y^2-x^2}\,\frac{\partial u}{\partial y} + z\,\frac{\partial u}{\partial z} = 0\,.$$

Let $\Omega \subseteq \mathbb{R}^3$ be a non-empty open set such that for each point (x,y,z) in Ω we have $x \neq 0$ and $y^2 \neq x^2$. The characteristic equation system is

$$\begin{aligned} y_1' &= \frac{t(t+2y_1)}{y_1^2-t^2}\,, \\ y_2' &= \frac{y_2}{t}\,. \end{aligned}$$

By the first equation we have

$$y_1'\,y_1^2 - y_1'\,t^2 - 2t\,y_1 - t^2 = 0\,,$$

$$\left(\frac{y_1^3}{3} - y_1\,t^2 - \frac{t^3}{3}\right)' = 0\,,$$

hence

$$\varphi_1(t,y_1,y_2) = \frac{y_1^3}{3} - y_1\,t^2 - \frac{t^3}{3}$$

is a first integral. The second equation implies $y_2(t) = C\cdot t$, consequently

$$\varphi_2(t,y_1,y_2) = \frac{y_2}{t}$$

is a first integral, too. As the rank of the matrix

$$\frac{\partial(\varphi_1,\varphi_2)}{\partial(t,y_1,y_2)} = \begin{pmatrix} -y_1t-t^2 & y_1^2-t^2 & 0 \\ -\frac{y_2}{t^2} & 0 & \frac{1}{t} \end{pmatrix}$$

is 2 everywhere, hence these first integrals are functionally independent. We conclude that the solution of the original equation has the form

$$u(x,y,z) = F\Big(\frac{y^3}{3} - y\,x^2 - \frac{x^3}{3}, \frac{z}{x}\Big)\,,$$

where $F:\mathbb{R}^2 \to \mathbb{R}$ is an arbitrary continuously differentiable function. This is what we call the general solution of our equation.

6.2 Problems

1. Find the general solution of the following partial differential equations:

   i)
$$\frac{\partial u}{\partial x} = \frac{\partial u}{\partial y}, \qquad u = u(x,y),$$

   ii)
$$x_1 \frac{\partial u}{\partial x_1} + x_2 \frac{\partial u}{\partial x_2} + \cdots + x_n \frac{\partial u}{\partial x_n} = 0, \qquad u = u(x_1, x_2, \ldots, x_n),$$

   iii)
$$y\frac{\partial u}{\partial x} - x\frac{\partial u}{\partial y} = 0, \qquad u = u(x,y).$$

2. Find the geometrical properties characterizing those surfaces satisfying the partial differential equation
$$x\frac{\partial u}{\partial x} + y\frac{\partial u}{\partial y} = 0\,.$$
   Which one is the surface corresponding to the solution u satisfying the additional condition $u(x,1) = x$?

3. Find the general solution of the partial differential equation
$$\sqrt{x}\,\frac{\partial f}{\partial x} + \sqrt{y}\,\frac{\partial f}{\partial y} + \sqrt{z}\,\frac{\partial f}{\partial z} = 0\,,$$
   and the solution satisfying the initial condition $f(1,y,z) = y - z$.

4. Find the solution of the partial differential equation
$$xz\,u_x + yz\,u_y - (x^2+y^2)\,u_z = 0$$
   satisfying $u(1,y,z) = 1 + z^2$.

5. Find the solution of the partial differential equation
$$y\,u_x - x\,u_y = 0$$
   passing through the curve $x = 1$, $y = s$, $u = s$.

6. Find the general solution of the following partial differential equations.

   i) $x\,u_x + y\,u_y = 0, \qquad u = u(x,y)\,,$
   ii) $u_x + z\,u_y = 0, \qquad u = u(x,y,z)\,,$
   iii) $xz\,u_x + yz\,u_y - (x^2+y^2)\,u_z = 0, \qquad u = u(x,y,z)\,,$
   iv) $x\,u_x + y^2\,u_y = 0, \qquad u = u(x,y)\,,$
   v) $x\,u_x + y\,u_y + (x^2+y^2)\,u_z = 0, \qquad u = u(x,y,z)\,.$

7. Solve the following Cauchy problems.

   i) $y\,u_x - x\,u_y = 0,\ x = \cos s,\ y = \sin s,\ u = 1\,,$
   ii) $u_x + u_y = 0,\ x = u = s,\ y = 0\,,$
   iii) $x\,u_x - y\,u_y = 0,\ x = y = u = s\,.$

6.3 Quasilinear partial differential equations

Let $\Omega \subseteq \mathbb{R}^n \times \mathbb{R}$ be a non-empty open set, $f : \Omega \to \mathbb{R}^n$ a continuous function, and $f_0 : \Omega \to \mathbb{R}$ a continuous function. The equation

$$\langle f(x,u), u'(x)\rangle = f_0(x,u) \tag{6.3}$$

is called *first order quasilinear partial differential equation.* We say that the function u *is a solution of* (6.3) *on* D, if $D \subseteq \mathbb{R}^n$ is a non-empty open set, $u : D \to \mathbb{R}$ is a continuously differentiable function, further for each x in D the point $\big(x, u(x)\big)$ belongs to Ω and (6.3) holds.

Let (x,u) be a point in Ω, and we define

$$\widetilde{f}(x,u) = \big(f(x,u), f_0(x,u)\big)\,.$$

The homogeneous linear partial differential equation

$$\langle \widetilde{f}(x,u), v'(x,u)\rangle = 0 \tag{6.4}$$

is called the *homogeneous linear partial differential equation corresponding to* (6.3), its characteristic equation system is called the characteristic equation system of (6.3), and a fundamental system of it is called a fundamental system of (6.3).

Theorem 6.3.1. *Let $\Omega_0 \subseteq \Omega$ and $D \subseteq \mathbb{R}^n$ be non-empty open sets, further let Φ be a solution of* (6.4) *on Ω_0 such that $\partial_{n+1}\Phi(x,u) \neq 0$, whenever (x,u) is in Ω_0. If $\varphi : D \to \mathbb{R}$ is a continuously differentiable function such that for each x in D the point $\big(x,\varphi(x)\big)$ belongs to Ω_0, and the function $x \mapsto \Phi\big(x,\varphi(x)\big)$ is constant on D, then φ is a solution of* (6.3) *on D.*

Proof. As Φ is a solution of (6.4) on Ω_0, hence for each (x,u) in Ω_0 we have

$$\sum_{i=1}^{n} \partial_i\Phi(x,u)\, f_i(x,u) + \partial_{n+1}\Phi(x,u)\, f_0(x,u) = 0\,.$$

The function $x \mapsto \Phi\big(x,\varphi(x)\big)$ is constant on D, hence its partial derivatives vanish:

$$\partial_i\Phi\big(x,\varphi(x)\big) + \partial_{n+1}\Phi\big(x,\varphi(x)\big)\,\partial_i\varphi(x) = 0\,,$$

whenever $i = 1, 2, \ldots, n$ and x is in D. From this we infer

$$\partial_i\varphi(x) = -\frac{\partial_i\Phi\big(x,\varphi(x)\big)}{\partial_{n+1}\Phi\big(x,\varphi(x)\big)}\,.$$

Substitution into (6.3) gives for each x in D

$$\langle f\big(x,\varphi(x)\big), \varphi'(x)\rangle = \sum_{i=1}^{n} f_i\big(x,\varphi(x)\big)\,\partial_i\varphi(x) = f_0\big(x,\varphi(x)\big)\,,$$

that is, φ is a solution of (6.3) on D. □

Theorem 6.3.2. *Let $\Omega_0 \subseteq \Omega$ be a non-empty open set, and let $f : \Omega \to \mathbb{R}^n$ be a continuous function which does not vanish at any point of Ω_0. Suppose that there exists a fundamental system of (6.3) on Ω_0. Assume, moreover, that $D \subseteq \mathbb{R}^n$ is a non-empty open set, and φ is a solution of (6.3) on D such that for each x in D the point $\big(x, \varphi(x)\big)$ lies in Ω_0. Then every point in D has a neighborhood $U \subseteq D$ such that there exists a solution Φ of (6.4) on Ω_0 for which the function $x \mapsto \Phi\big(x, \varphi(x)\big)$ is constant on U.*

Proof. We introduce the functions on D defined by

$$g_i(x) = f_i\big(x, \varphi(x)\big), \qquad \psi_i(x) = \Phi_i\big(x, \varphi(x)\big),$$

where $f = (f_1, f_2, \ldots, f_n)$, and $\Phi_1, \Phi_2, \ldots, \Phi_n$ is a fundamental system of (6.3) on Ω_0. Then for each x in D we have

$$\sum_{k=1}^{n} \partial_k \psi_i(x)\, g_k(x)$$

$$= \sum_{k=1}^{n} \big(\partial_k \Phi_i(x, \varphi(x)) + \partial_{n+1}\Phi_i(x, \varphi(x))\, \partial_k \varphi(x)\big) f_k\big(x, \varphi(x)\big)$$

$$= \sum_{k=1}^{n} \big(\partial_k \Phi_i(x, \varphi(x)) + \partial_{n+1}\Phi_i(x, \varphi(x)) \cdot \sum_{k=1}^{n} \partial_k \varphi(x)\big) f_k\big(x, \varphi(x)\big)$$

$$= \sum_{k=1}^{n} \big(\partial_k \Phi_i(x, \varphi(x)) + \partial_{n+1}\Phi_i(x, \varphi(x)) \cdot f_0\big(x, \varphi(x)\big) = 0,$$

for $i = 1, 2, \ldots, n$. Hence the functions $\psi_1, \psi_2, \ldots, \psi_n$ are solutions of the homogeneous linear partial differential equation

$$\langle g(x), u'(x) \rangle = 0$$

on D, where $g = (g_1, g_2, \ldots, g_n)$. As g does not vanish at the point of D, by Theorem 6.1.1, the functions $\psi_1, \psi_2, \ldots, \psi_n$ are functionally independent on D. Hence for every point of D there is a neighborhood of this point $U \subseteq D$ and a continuously differentiable function $F : \mathbb{R}^n \to \mathbb{R}$ such that the function $x \mapsto F\big(\psi_1(x), \psi_2(x), \ldots, \psi_n(x)\big)$ is constant on U. Now we define $\Phi = F \circ (\Phi_1, \Phi_2, \ldots, \Phi_n)$, then obviously Φ is a solution of (6.4) on Ω_0, and, by the definition of the functions ψ_i, the function $x \mapsto \Phi\big(x, \varphi(x)\big)$ is constant on U. $\square$

Using Theorem 6.3.1 we can solve equation (6.3) in the following way: we express the variable u from the equation

$$F\big(\Phi_1(x,u), \Phi_2(x,u), \ldots, \Phi_n(x,u)\big) = 0\,.$$

Here $\Phi_1, \Phi_2, \ldots, \Phi_n$ is a fundamental system of (6.3), and F is an arbitrary continuously differentiable function. Theorem 6.3.2 states that every solution of (6.3) can be obtained in this manner. Hence the implicit equation above for u is called the *general solution of* (6.3).

As an illustration we solve the partial differential equation

$$x\,\frac{\partial u}{\partial x} + y\,\frac{\partial u}{\partial y} = u\,.$$

Let $D \subseteq R^2$ be a non-empty open set such that $\neq 0$, whenever (x,y) is in D. The corresponding homogeneous equation is

$$x\,\frac{\partial v}{\partial x} + y\,\frac{\partial v}{\partial y} + u\,\frac{\partial v}{\partial u} = 0\,.$$

The characteristic equation system is

$$y_1' = \frac{y_1}{t}\,,$$
$$y_2' = \frac{y_2}{t}\,,$$

which implies $y_1(t) = C_1 \cdot t$, $y_2(t) = C_2 \cdot t$. Hence the functions

$$\Phi_1(t, y_1, y_2) = \frac{y_1}{t}, \qquad \Phi_2(t, y_1, y_2) = \frac{y_2}{t}$$

are independent first integrals which form a fundamental system of the original equation. Consequently, the general solution is given by the implicit form

$$F\left(\frac{y}{x}, \frac{u}{x}\right) = 0\,,$$

where F is an arbitrary continuously differentiable function. If $\partial_2 F \neq 0$, then u can be expressed in the explicit form

$$u(x,y) = x \cdot f\left(\frac{y}{x}\right),$$

where f is an arbitrary continuously differentiable function.

6.4 Problems

1. Solve the following problems:

   i) $(y+z+u)\,u_x + (z+u+x)\,u_y + (u+x+y)\,u_z = x+y+z$,
   $u = u(x,y,z)$,

   ii)
   $$a\,\frac{\partial u}{\partial x} + b\,\frac{\partial u}{\partial y} + c\,\frac{\partial u}{\partial z} = xyz, \quad a,b,c \text{ are constants},$$

   iii) $(y^3x - 2x^4)\,u_x + (2y^4 - x^3y)\,u_y = 9u(x^3 - y^3)$, $\quad u = u(x,y)$,

   iv) $z(x+z)\,z_x - y(y+z)\,z_y = 0$, $\quad z = z(x,y)$, $\quad z(1,y) = \sqrt{y}$.

2. Solve the following problems.

   i) $x\,u_x + y\,u_y = au$, $\quad u = u(x,y)$, $\quad a$ is a constant,

   ii) $y\,u_x - x\,u_y = 2xyu$, $\quad u = u(x,y)$,

   iii) $x^2\,u_x - xy\,u_y + y^2 = 0$, $\quad u = u(x,y)$,

   iv) $x_1\,u_{x_1} + x_2\,u_{x_2} + \cdots + x_n\,u_{x_n} = a\,u$,
   $u = u(x_1, x_2, \ldots, x_n)$, a is a constant,

   v) $u_x + a\,u_y + b\,u_z = xyz$,
   $u = u(x,y,z)$, $\quad a,b$ are constants,

   vi) $(au - by)\,u_x + (bx - cu)\,u_y = cy - ax$,
   $u = u(x,y)$, a,b,c are constants,

   vii) Find those surfaces intersecting the cones $xy = az^2$ at right angles.

3. Solve the following Cauchy problems.

   i) $u_x + u_y + u + x + y + 2 = 0$, $\quad x = s$, $\quad y = 0$, $\quad u = s$,

   ii) $y\,u_x - x\,u_y = x^3y + xy^3$, $\quad x = 0$, $\quad y = s$, $\quad u = s$,

   iii) $x\,u_x + y\,u_y = u$, $\quad x = \cos s$, $\quad y = \sin s$, $\quad u = 1$,

   iv) $u_x + y\,u_y - z\,u_z = z(1-x)$, $\quad u(0,y,z) = yz$,

   v) $xu\,u_x + yu\,u_y = x^2 + y^2 + u^2$, $\quad u(1,y) = y^2$,

   vi)
   $$u\,u_x + u_y = 1, \quad x = \frac{s^2}{2}, \quad y = s, \quad u = s.$$

Chapter 7

THEORY OF CHARACTERISTICS

7.1 First order partial differential equations

In this chapter n denotes an integer greater than 1. Let $\Omega \subseteq \mathbb{R}^n \times \mathbb{R} \times \mathbb{R}^n$ be a non-empty open set, and let $F : \Omega \to \mathbb{R}$ be a continuous function. Then the equation

$$F\big(x, u(x), u'(x)\big) = 0 \tag{7.1}$$

is called *first order partial differential equation.* We say that the function u is *a solution of equation* (7.1) *on* T, if $T \subseteq \mathbb{R}^n$ is a non-empty open set, $u : T \to \mathbb{R}$ is a continuously differentiable function, for each x in T the point $\big(x, u(x), u'(x)\big)$ belongs to Ω, further (7.1) holds.

From now on we always suppose that F is continuously differentiable.

The system of ordinary differential equations

$$\begin{aligned} x' &= \partial_3 F(x, u, p) \\ u' &= \langle \partial_3 F(x, u, p), p \rangle \\ p' &= -\big[\partial_1 F(x, u, p) + \partial_2 F(x, u, p) \cdot p\big] \end{aligned} \tag{7.2}$$

is called the *characteristic differential equation system* of (7.1). Hence this is an ordinary first order system consisting of $n + 1 + n = 2n + 1$ equations for the unknown functions $x = (x_1, x_2, \ldots, x_n)$, u and $p = (p_1, p_2, \ldots, p_n)$, whose number is $2n + 1$, too. We recall that in the $n + 1$-th equation $\langle , \rangle$ denotes the euclidean inner product in $\mathbb{R}^n$. If (x, u, p) is a solution of this system, then (x, u) is called a *characteristic curve.* Hence characteristic curves are determined by the first $n + 1$ equations which also include the components $p_1, p_2, \ldots, p_n$ of the auxiliary function p, too.

We say that the triplet (x, u, p) of functions forms a *strip* on the non-empty open interval $I \subseteq \mathbb{R}$, if $x : I \to \mathbb{R}^n$, $u : I \to \mathbb{R}$ are continuously differentiable functions, $p : I \to \mathbb{R}^n$ is a continuous function, further the so-called *strip condition*

$$u' = \langle x', p \rangle$$

holds. This has the following simple geometrical interpretation: to each point $\big(x(t), u(t)\big)$ of the curve (x, u) we assign a plane element with normal vector $\big(p(t), -1\big)$. By the strip condition, the tangent vector of the curve (x, u) at each point is perpendicular to this normal vector, that is, it is included in the plane assigned to the point.

By the first $n+1$ equations of the characteristic equation system of (7.1) each solution of the characteristic system is a strip. A solution (x, u, p) is called a *characteristic strip*, if the function $t \mapsto F\big(x(t), u(t), p(t)\big)$ vanishes identically. It is easy to see that this function is always constant, hence for (x, u, p) is a characteristic strip it is necessary and sufficient that the function $t \mapsto F\big(x(t), u(t), p(t)\big)$ vanishes at some point. Indeed, we have

$$\frac{d}{dt} F(x, u, p) = \langle F_x, x' \rangle + F_u \cdot u' + \langle F_p, p' \rangle$$

$$= \langle F_x, F_p \rangle + F_u \cdot \langle F_p, p \rangle - \langle F_p, F_x \rangle - F_u \cdot \langle F_p, p \rangle = 0 \,.$$

Let $T \subseteq \mathbb{R}^n$ be a non-empty open set, and let $\Phi : T \to \mathbb{R}$ be a continuously differentiable function. The triplet (x_0, u_0, p_0) is called a *surface element* of the surface Φ, if x_0 is a point in T, $u_0 = \Phi(x_0)$, and $p_0 = \Phi'(x_0)$. We say that the strip (x, u, p), defined on I, *fits to the surface element* (x_0, u_0, p_0), if for some t_0 in I we have $x_0 = x(t_0)$, $u_0 = u(t_0)$, $p_0 = p(t_0)$.

Theorem 7.1.1. *(Fundamental Theorem) Let $T \subseteq \mathbb{R}^n$ be a non-empty open set. The twice continuously differentiable function $\Phi : T \to \mathbb{R}$ is a solution of* (7.1) *on T if and only if for each surface element of Φ there is a characteristic strip of* (7.1) *which fits to it.*

Proof. Let Φ be a solution of (7.1) on T, and let (x_0, u_0, p_0) be an arbitrary surface element of Φ. By the existence theorem 1.5.1 of Peano, the system of ordinary differential equations

$$x' = \partial_3 F(x, \Phi \circ x, \Phi' \circ x)$$

has a solution x such that $x(t_0) = x_0$ holds for some real t_0. Let $u = \Phi \circ x$, $p = \Phi' \circ x$. We have

$$x' = \partial_3 F(x, u, p) \,,$$

$$u' = \langle \Phi' \circ x, x' \rangle = \langle x', p \rangle = \langle \partial_3 F(x, u, p), p \rangle ,$$

and

$$p' = (\Phi'' \circ x) \cdot x' .$$

On the other hand, for each point x in T we obtain

$$F\big(x, \Phi(x), \Phi'(x)\big) = 0 ,$$

hence, by differentiation with respect to x, it follows

$$\partial_1 F + \partial_2 F \cdot \Phi' + \Phi'' \cdot \partial_3 F = 0 ,$$

and, by the substitution $x = x(t)$, we get

$$\partial_1 F(x, u, p) + \partial_2 F(x, u, p) \cdot p + (\Phi'' \circ x) \cdot x' = 0 ,$$

which gives

$$p' = (\Phi'' \circ x) \cdot x' = -\partial_1 F(x, u, p) - \partial_2 F(x, u, p) \cdot p .$$

Finally, we conclude that

$$F\big(x(t_0), u(t_0), p(t_0)\big) = F(x_0, u_0, p_0) = F\big(x_0, \Phi(x_0), \Phi'(x_0)\big) = 0 ,$$

which means that (x, u, p) is a characteristic strip, and it fits to the surface element (x_0, u_0, p_0).

Conversely, assume that the condition of the theorem holds, and x_0 is an arbitrary point in T. Let $u_0 = \Phi(x_0)$, $p_0 = \Phi'(x_0)$, and let (x, u, p) be a characteristic strip which fits to the surface element (x_0, u_0, p_0): $x_0 = x(t_0)$, $u_0 = u(t_0)$, $p_0 = p(t_0)$. Then we have

$$F\big(x_0, \Phi(x_0), \Phi'(x_0)\big) = F(x_0, u_0, p_0) = F\big(x(t_0), u(t_0), p(t_0)\big) = 0 ,$$

and as x_0 is arbitrary, we have proved that Φ is a solution of (7.1) on T. □

7.2 Problems

Find the characteristic differential equation system of the given partial differential equations:

1. $x u_x + y u_y - u_x u_y = 0$
2. $a u_x + b u_y + c u_z = xyz$
3. $u_x u_y = u$
4. $u_x^2 + u_y^2 = 1$
5. $x u_x + y u_y = u - xy$

7.3 Cauchy problem for first order equations

Let $\Omega \subseteq \mathbb{R}^n \times \mathbb{R} \times \mathbb{R}^n$ be a non-empty open set, $F : \Omega \to \mathbb{R}$ a continuous function, let $D \subseteq \mathbb{R}^{n-1}$ be a non-empty open set, further let $\varphi : D \to \mathbb{R}^n$, $\psi : D \to \mathbb{R}$ be continuous functions. The equation

$$u\big(\varphi(s)\big) = \psi(s) \tag{7.3}$$

is called an *initial condition* for (7.1), and the system of equations (7.1), (7.3) is called an *initial value problem*, or *Cauchy problem* for the first order partial differential equation (7.1). We say that u is a solution of the Cauchy problem (7.1), (7.3) on T, if u is a solution of (7.1) on T such that for each s in D the point $\varphi(s)$ belongs to T, and $u\big(\varphi(s)\big) = \psi(s)$ holds. Geometrically this means that the surface u fits to that part of the initial object (φ, ψ) which is "above" its domain of definition. For instance, for $n = 2$ we require that the solution surface fits to a space curve. From the geometrical point of view the solution of the Cauchy problem (7.1), (7.3) means the following procedure: first we complete the initial object to an initial strip, then we fit a characteristic strip to each strip element of this initial strip. Finally, we try to identify the envelop surface of this family of characteristic strips. For the uniqueness it is obviously necessary that the initial strip is non-characteristic, further that the family of characteristic strips determined above has a continuously differentiable envelop surface. The assumptions in the following theorem are to guarantee these necessary conditions. The proof of the theorem is constructive as it provides an effective method to find the solution of the Cauchy problem in particular cases. We call the reader's attention that the given conditions assure uniqueness only in the set of twice continuously differentiable functions.

Theorem 7.3.1. *(Existence and Uniqueness Theorem) Let F, φ, ψ in the Cauchy problem* (7.1), (7.3) *be twice continuously differentiable functions, further let s_0 be a point in D, p_0 a point in $\mathbb{R}^n$, and $x_0 = \varphi(s_0)$, $u_0 = \psi(s_0)$. If the conditions*

$$F(x_0, u_0, p_0) = 0\,, \tag{7.4}$$

$$\varphi'(s_0) \cdot p_0 = \psi'(s_0)\,, \tag{7.5}$$

$$\det\big(\partial_3 F(x_0, u_0, p_0), \varphi'(s_0)\big) \neq 0\,, \tag{7.6}$$

hold, then the point x_0 has a neighborhood on which the Cauchy problem (7.1), (7.3) *has exactly one twice continuously differentiable solution.*

Proof. We split the proof into five steps.

Step One: By the Implicit Function Theorem, there exists a neighborhood $U_1(s_0) \subseteq D$ of the point s_0, and there exists a unique continuously differentiable function $\chi : U_1(s_0) \to \mathbb{R}^n$ such that $\chi(s_0) = p_0$, and for each s in $U_1(s_0)$ we have

$$F\big(\varphi(s), \psi(s), \chi(s)\big) = 0\,,$$
$$\varphi'(s) \cdot \chi(s) = \psi'(s)\,.$$

In other words, the variable p can be expressed in some neighborhood of s_0 in a continuously differentiable way from the system of equations

$$F\big(\varphi(s), \psi(s), p\big) = 0\,,$$
$$\varphi'(s) \cdot p = \psi'(s)$$

in terms of the variable s. This possibility is guaranteed by the condition in (7.6).

In this way the initial object is extended to an initial strip. In the next step we fit a characteristic strip to each strip element of this initial strip.

Step Two: Let (x, u, p) denote the characteristic function of the characteristic system of differential equations (7.2), and let

$$X(t, s) = x\big(t, t_0, \varphi(s), \psi(s), \chi(s)\big)\,,$$

$$U(t, s) = u\big(t, t_0, \varphi(s), \psi(s), \chi(s)\big)\,,$$

$$P(t, s) = p\big(t, t_0, \varphi(s), \psi(s), \chi(s)\big)\,.$$

In other words, the function $t \mapsto \big(X(t, s), U(t, s), P(t, s)\big)$ is the solution of the characteristic system of differential equations with the initial conditions

$$x(t_0) = \varphi(s)\,,$$

$$u(t_0) = \psi(s)\,,$$

$$p(t_0) = \chi(s)\,.$$

The characteristic function exists, as the function F is twice continuously differentiable. The same condition implies that there exist neighborhoods $U_1(t_0)$, respectively $U_2(s_0) \subseteq U_1(s_0)$ of t_0, respectively s_0 such that the functions X, U, P are continuously differentiable on $U_1(t_0) \times U_2(s_0)$. Hence for each s in $U_1(s_0)$ we have

$$X(t_0, s) = \varphi(s), \qquad U(t_0, s) = \psi(s), \qquad P(t_0, s) = \chi(s)\,,$$

moreover, for each s in $U_2(s_0)$ the function $t \mapsto \big(X(t,s), U(t,s), P(t,s)\big)$ is a characteristic strip of (7.1) on $U_1(t_0)$.

Step Three: Now we show that the family of characteristic strips we have obtained has an envelop surface Φ. We know that X is continuously differentiable on $U_1(t_0) \times U_2(s_0)$, and

$$X'(t_0, s_0) = \big(\partial_1 X(t_0, s_0), \partial_2 X(t_0, s_0)\big).$$

We show that this matrix is non-singular. Indeed, this is exactly the matrix occurring in condition (7.6), because we have

$$\partial_1 X(t,s) = \partial_3 F\big(X(t,s), U(t,s), P(t,s)\big),$$

consequently

$$\partial_1 X(t_0, s_0) = \partial_3 F(x_0, u_0, p_0),$$

and

$$\partial_2 X(t_0, s_0) = \varphi'(s_0).$$

As $X'(t_0, s_0)$ is non-singular, hence, by the Inverse Function Theorem, there exists a neighborhood $U_2(t_0) \subseteq U_1(t_0)$, respectively $U_3(s_0) \subseteq U_2(s_0)$ of t_0, respectively s_0 such that X is one-to-one on $U_2(t_0) \times U_3(s_0)$. The inverse of the restriction of X to $U_2(t_0) \times U_3(s_0)$ is denoted by Y. Then Y is continuously differentiable in some neighborhood $U(x_0)$ of x_0, and

$$Y'(x_0) = \big[X'(t_0, s_0)\big]^{-1}.$$

For each x in $U(x_0)$ we let $\Phi(x) = U\big(Y(x)\big)$. Obviously, Φ is continuously differentiable.

Step Four: We show that Φ is a twice continuously differentiable solution of the Cauchy problem (7.1), (7.3) on $U(x_0)$.

Let $G = Y\big(U(x_0)\big)$, then $G \subseteq U_2(t_0) \times U_3(s_0)$ is a neighborhood of (t_0, s_0). If s is a point in D, and $\varphi(s)$ belongs to $U(x_0)$, then we infer

$$\Phi\big(\varphi(s)\big) = U\big[Y\big(\varphi(s)\big)\big] = U\big[Y\big(X(t_0, s)\big)\big],$$

$$\Phi'\big[X(t,s)\big] \cdot X'(t,s) = U'(t,s), \qquad \Phi'\big[X(t,s)\big] = U'(t,s) \cdot X'(t,s)^{-1}.$$

We want to show that $\Phi'\big[X(t,s)\big] = P(t,s)$, that is

$$U'(t,s) = P(t,s) \cdot X'(t,s).$$

This matrix equation is equivalent to the equations

$$\partial_1 U(t,s) = \langle P(t,s), \partial_1 X(t,s)\rangle,$$

$$\partial_2 U(t,s) = \partial_2 X(t,s) \cdot P(t,s)\,.$$

The first equation is obvious, as for each s in $U_2(s_0)$ the function

$$t \mapsto \big(X(t,s), U(t,s), P(t,s)\big)$$

is a characteristic strip of (7.1). To prove the second equation first we differentiate the equations

$$U_t(t,s) = P(t,s) \cdot X_t(t,s)\,,$$

and

$$F\big(X(t,s), U(t,s), P(t,s)\big) = 0$$

with respect to s. Then we obtain

$$U_{t,s} = P_s \cdot X_t + X_{t,s} \cdot P\,,$$

$$X_s \cdot F_x + U_s \cdot F_u + P_s \cdot F_p = 0\,.$$

By the equation $F_x = -F_u \cdot P - P_t$, we deduce from the second equation the following one:

$$-F_u(X_s \cdot P - U_s) - X_s \cdot P_t + U_{t,s} - X_u \cdot P = 0\,,$$

or, by reordering,

$$\frac{d}{dt}(U_s - X_s \cdot P) = -F_u \cdot (U_s - X_s \cdot P)\,.$$

For each value of the parameter s this is an ordinary first order differential equation for the function $t \mapsto U_s(t,s) - X_s(t,s) \cdot P(t,s)$, and, by

$$U_s(t_0,s) - X_s(t_0,s) \cdot P(t_0,s) = \psi'(s) - \varphi'(s) \cdot \chi(s) = 0\,,$$

the uniqueness theorem implies that $U_s - X_s \cdot P = 0$ on the whole interval $U_2(t_0)$. This was to be proved, hence we have

$$\Phi'\big[X(t,s)\big] = P(t,s)\,.$$

Consequently, for each x in $U(x_0)$ we have $\Phi'(x) = P\big[Y(x)\big]$, and

$$F\big(x, \Phi(x), \Phi'(x)\big) = F\big(X(t,s), U(t,s), P(t,s)\big) = 0\,.$$

This shows that Φ is a solution of the Cauchy problem (7.1), (7.3). On the other hand, as P and Y are continuously differentiable, and, by the above equation, $\Phi'(x) = P\big[Y(x)\big]$, hence our Φ is twice continuously differentiable.

Step Five: Now we show that the solution obtained above is unique in a neighborhood of x_0 in the collection of twice continuously differentiable functions. Indeed, $\Psi : U(x_0) \to \mathbb{R}$ is a twice continuously differentiable solution of the Cauchy problem (7.1), (7.3), then for each s in D we have the equation $\Psi\big(\varphi(s)\big) = \psi(s)$, and $\varphi(s)$ belongs to $U(x_0)$ which implies $\varphi'(s) \cdot \Psi'\big(\varphi(s)\big) = \psi'(s)$, further we have

$$F\big(\varphi(s), \psi(s), \Psi'(s)\big) = 0 .$$

By the uniqueness statement of the Implicit Function Theorem, we deduce $\chi(s) = \Psi'(s)$. For each s in $U_2(s_0)$ we consider the Cauchy problem

$$\begin{aligned} x' &= \partial_3 F(x, \Psi \circ x, \Psi' \circ x) , \\ x(t_0) &= \varphi(s) , \end{aligned}$$

where t_0 is an arbitrary real number. Under the condition for F this has a unique solution. We denote the solution by $t \mapsto Z(t, s)$, further let

$$\begin{aligned} V(t, s) &= \Psi\big(Z(t, s)\big) , \\ Q(t, s) &= \Psi'\big(Z(t, s)\big) . \end{aligned}$$

Then it is easy to check that the function $t \mapsto \big(Z(t, s), V(t, s), Q(t, s)\big)$ is a solution of the characteristic system of differential equations (7.2). On the other hand, we have

$$Z(t_0, s) = \varphi(s) ,$$

$$V(t_0, s) = \Psi\big(Z(t_0, s)\big) = \Psi\big(\varphi(s)\big) = \psi(s) ,$$

$$Q(t_0, s) = \Psi'\big(Z(t_0, s)\big) = \Psi'\big(\varphi(s)\big) = \chi(s) ,$$

hence, by the uniqueness theorem,

$$Z(t, s) = X(t, s), \qquad V(t, s) = U(t, s), \qquad Q(t, s) = P(t, s)$$

holds in some neighborhood of (t_0, s_0). Consequently, in the corresponding neighborhood of x_0 we obtain

$$\Psi(x) = \Psi\big(Z(t, s)\big) = V(t, s) = U(t, s) = \Psi\big(X(t, s)\big) = \Phi(x) ,$$

which proves the uniqueness and the theorem is proved. □

Suppose that the conditions of the previous theorem are satisfied for the Cauchy problem (7.1), (7.3). Then we can find the solution of the Cauchy problem in the following manner: first by solving the system of algebraic equations

$$\begin{aligned} F\big(\varphi(s), \psi(s), \chi(s)\big) &= 0 , \\ \varphi'(s) \cdot \chi(s) &= \psi'(s) \end{aligned}$$

for $\chi(s)$ we extend the initial object (φ, ψ) to an initial strip: (φ, ψ, χ). Then we determine the solution of the characteristic system of differential equations satisfying the initial conditions

$$X(t_0, s) = \varphi(s), \qquad U(t_0, s) = \psi(s), \qquad P(t_0, s) = \chi(s)\,.$$

The next step is to express the variable (t, s) from the system of equations $x = X(t, s)$ in terms of x, and substitute it into $u = U(t, s)$. Finally, we get the solution in the form

$$u(x) = U\big(t(x), s(x)\big)\,.$$

As an illustration we solve the first order partial differential equation

$$\left(\frac{\partial u}{\partial x}\right)^2 + 2y^2 \cdot \frac{\partial u}{\partial y} = 1$$

with the initial condition

$$u(s, 1) = \frac{s^2}{2}\,.$$

Using the notation of the theorem we let $x \sim (x, y)$, $p \sim (p, q)$, then

$$F(x, y, u, p, q) = p^2 + 2y^2\, q - 1\,,$$

$$\varphi(s) = (s, 1), \qquad \psi(s) = \frac{s^2}{2}\,.$$

Let $s_0 = 0$, then we have $x_0 = 0$, $y_0 = 1$, $u_0 = 0$. The values p_0, q_0 we obtain from the system of equations

$$p_0^2 + 2q_0 - 1 = 0\,,$$
$$\langle (1, 0), (p_0, q_0) \rangle = 0$$

in the form $p_0 = 0$, $q_0 = \frac{1}{2}$. The matrix in condition (7.6) is

$$\begin{pmatrix} 0 & 1 \\ 2 & 0 \end{pmatrix},$$

it is non-singular, hence all conditions of the theorem are satisfied.

Step One: We solve for (p, q) the system of equations

$$p^2 + 2q - 1 = 0\,,$$
$$\langle (1, 0), (p, q) \rangle = s\,.$$

The solution is $p = s$, $q = \frac{1-s^2}{2}$. Hence $\chi(s) = \left(s, \frac{1-s^2}{2}\right)$, and the initial strip is

$$(s, 1), \frac{s^2}{2}, \left(s, \frac{1-s^2}{2}\right).$$

Step Two: We solve the characteristic system of differential equations

$$\begin{aligned} x' &= 2p\,, \\ y' &= 2y^2\,, \\ u' &= 2p^2 + 2y^2 q\,, \\ p' &= 0\,, \\ q' &= -4yq \end{aligned}$$

with the initial conditions

$$\begin{aligned} x(0) &= s\,, \\ y(0) &= 1\,, \\ u(0) &= \frac{s^2}{2}\,, \\ p(0) &= s\,, \\ q(0) &= \frac{1-s^2}{2}\,. \end{aligned}$$

The solution is

$$\begin{aligned} X(t,s) &= 2st + s\,, \\ Y(t,s) &= \frac{1}{1-2t}\,, \\ U(t,s) &= \frac{s^2}{2}(2t+1) + t\,, \\ P(t,s) &= s\,, \\ Q(t,s) &= \frac{1-s^2}{2}(1-2t)^2\,. \end{aligned}$$

Step Three: From the system of equations

$$\begin{aligned} x &= 2st + s\,, \\ y &= \frac{1}{1-2t} \end{aligned}$$

we express t and s in terms of x and y. We obtain

$$t = \frac{1}{2} - \frac{1}{2y}\,, \qquad s = \frac{xy}{2y-1}\,.$$

Finally, we substitute this into the formula for u:

$$u(x,y) = U(t,s) = \frac{1}{2}\left(\frac{x^2y}{2y-1} + \frac{y-1}{y}\right).$$

This function u is the solution of the Cauchy problem which is unique in the family of twice continuously differentiable functions.

7.4 Problems

Solve the following Cauchy problems:

1. $x\,u_x + y\,u_y - u_x\,u_y = 0, \quad u(0,y) = y\,,$
2. $u_x\,u_y = u, \quad u(0,s) = s^2\,,$
3. $$\frac{\partial u}{\partial x} - \left(\frac{\partial u}{\partial y}\right)^2 = 0, \quad u(s,s^2) = s^3,$$
4. $u_x^2 + u_y^2 = 1, \quad x = s, \quad y = s, \quad u = s\,,$
5. $u_x\,u_y = \frac{1}{2}, \quad u(0,y) = y\,,$
6. $u_x^2 + u_y^2 = 1, \quad x = s, \quad y = -s, \quad u = 1\,.$

7.5 Special Cauchy problem for first order partial differential equation

Let $\Omega \subseteq \mathbb{R}^{2n+2}$ and $T \subseteq \mathbb{R}^n$ be non-empty open sets, let $f : \Omega \to \mathbb{R}$ and $g : T \to \mathbb{R}$ be continuous functions, further let t_0 be a real number. Then the system of equations

$$u_t + f(t,x,u,u') = 0\,, \tag{7.7}$$

$$u(t_0,x) = g(x) \tag{7.8}$$

is called *special Cauchy problem* for the first order partial differential equation (7.7). Clearly, (7.7) is a special case of (7.1), and (7.8) is a special case of (7.3). Accordingly, the function Φ is a solution of the special Cauchy problem (7.7), (7.8), if and only if $\Phi : D \to \mathbb{R}$ is a continuously differentiable function, $D \subseteq \mathbb{R} \times \mathbb{R}^n$ is a non-empty open set, there exists an x in $\mathbb{R}^n$ such that (t_0,x) belongs to D, further for each (t,x) in D we have

$$\partial_1\,\Phi(t,x) + f\big(t,x,\Phi(t,x),\partial_2\Phi(t,x),\ldots,\partial_{n+1}\Phi(t,x)\big) = 0\,,$$

moreover the equation $\Phi(t_0,x) = g(x)$ holds whenever (t_0,x) is in D.

Theorem 7.5.1. *Let $\Omega \subseteq \mathbb{R}^{2n+2}$ and $T \subseteq \mathbb{R}^n$ be non-empty open sets, let $f : \Omega \to \mathbb{R}$ and $g : T \to \mathbb{R}$ be twice continuously differentiable functions, let t_0 be a real number, and let x_0 an arbitrary point in $\mathbb{R}^n$. Suppose that $u_0 = g(x_0)$, $p_0 = g'(x_0)$, and (t_0, x_0, u_0, p_0) belongs to Ω. Then the point (t_0, x_0) has a neighborhood such that the special Cauchy problem (7.7), (7.8) has exactly one twice continuously differentiable solution.*

Proof. We show that the conditions of our present theorem imply the conditions in Theorem 7.3.1.

Using the notation in Theorem 7.3.1 we have here $n \sim n+1$, $x \sim (t, x)$, $u \sim u$, $p \sim (q, p)$, $s \sim x$, $s_0 \sim x_0 = \varphi(s_0)$, $u_0 = \psi(s_0) \sim g(x_0)$, further

$$F(t, x, u, q, p) = q + f(t, x, u, p),$$

$$\varphi(x) = (t_0, x), \qquad \psi(x) = g(x),$$

and finally

$$q_0 = -f\big(t_0, x_0, g(x_0), g'(x_0)\big), \qquad p_0 \sim \big(q_0, g'(x_0)\big).$$

Then φ and ψ are twice continuously differentiable, and we have

$$q_0 + f\big(t_0, x_0, g(x_0, g'(x_0)\big) = 0,$$

moreover

$$\begin{pmatrix} 0 & 1 & 0 & \dots & 0 \\ 0 & 0 & 1 & \dots & 0 \\ \dots & \dots & \dots & \dots & \dots \\ 0 & 0 & 0 & \dots & 1 \end{pmatrix} \cdot \begin{pmatrix} q_0 \\ \partial_1 g(x_0) \\ \dots \\ \partial_n g(x_0) \end{pmatrix} - g'(x_0) = 0,$$

and finally

$$\det \begin{pmatrix} 1 & 0 & \dots & \dots & 0 \\ f_{p_1} & 1 & 0 & \dots & 0 \\ f_{p_2} & 0 & 1 & \dots & 0 \\ \dots & & & & \\ f_{p_n} & 0 & 0 & \dots & 1 \end{pmatrix} = 1 \neq 0.$$

This completes the proof of the theorem. □

We note that by specializing rather the proof of Theorem 7.3.1, than the statement of it we can get a stronger result, in particular, a global existence and uniqueness theorem can be obtained.

By the above consideration, the geometric meaning of the general Cauchy problem for the quasilinear partial differential equation (6.3) is to find a solution Φ of equation (6.3) such that, as a surface, it fits to an $n-1$ dimensional object lying in the domain Ω of the partial differential equation. Such an object can be described by some continuously differentiable functions

$$\varphi : T \to \mathbb{R}^n, \qquad \psi : T \to \mathbb{R},$$

where $T \subseteq \mathbb{R}^{n-1}$ is a non-empty open set. The condition that the surface Φ fits to the object defined by the pair of functions φ, ψ means that for each point s in T the function Φ is defined at s, further we have

$$\Phi\big(\varphi(s)\big) = \psi(s).$$

To solve the Cauchy problem we assume that we have determined a fundamental system $\Phi_1, \Phi_2, \ldots, \Phi_n$ of (6.3) on some non-empty open set $D_0 \subseteq \Omega$. By the general theory, we know that locally there exists a continuously differentiable function F such that we have

$$F\big(\Phi_1(x, \Phi(x)), \Phi_2(x, \Phi(x)), \ldots, \Phi_n(x, \Phi(x))\big) = 0,$$

that is, Φ can be obtained by expressing u from the equation

$$F\big(\Phi_1(x, u), \Phi_2(x, u), \ldots, \Phi_n(x, u)\big) = 0.$$

The question is how to choose F in order that Φ fits to the given object? To find F we have to eliminate the variable s from the equation system

$$\begin{aligned} \xi_1 &= \Phi_1\big(\varphi(s), \psi(s)\big), \\ \xi_2 &= \Phi_2\big(\varphi(s), \psi(s)\big), \\ &\ldots \\ \xi_n &= \Phi_n\big(\varphi(s), \psi(s)\big), \end{aligned}$$

then the function F occuring in

$$F(\xi_1, \xi_2, \ldots, \xi_n) = 0$$

will be the desired one.

As an illustration we solve the following Cauchy problem:

$$\frac{1}{x}\frac{\partial u}{\partial x} + \frac{1}{y}\frac{\partial u}{\partial y} + \frac{1}{u} = 0,$$

$$x = a\cos s,$$

$$y = a\sin s,$$

$$u = b.$$

The corresponding homogeneous equation is

$$\frac{1}{x}\frac{\partial v}{\partial x} + \frac{1}{y}\frac{\partial v}{\partial y} - \frac{1}{u}\frac{\partial v}{\partial u} = 0\,,$$

and the characteristic equation system is

$$\begin{aligned} x' &= \frac{1}{x}\,, \\ y' &= \frac{1}{y}\,, \\ u' &= -\frac{1}{u}\,. \end{aligned}$$

We have the two independent first integrals:

$$\Phi_1(x, y, u) = x^2 - y^2, \qquad \Phi_2(x, y, u) = x^2 + u^2\,,$$

hence the system of equations we have to eliminate s from has the form

$$\begin{aligned} \xi_1 &= a^2\cos^2 s - a^2\sin^2 s\,, \\ \xi_2 &= a^2\cos^2 s + b^2\,. \end{aligned}$$

By elimination we obtain

$$\xi_1 - 2\xi_2 + 2b^2 + a^2 = 0\,,$$

which gives

$$F(\xi_1, \xi_2) = \xi_1 - 2\xi_2 + 2b^2 + a^2\,,$$

hence the solution of the Cauchy problem is given by the implicit function

$$x^2 - y^2 - 2x^2 - 2u^2 + 2b^2 + a^2 = 0$$

in the form

$$u^2 = \frac{a^2 - x^2 - y^2}{2} + b^2\,.$$

In the particular cases it can be very complicated to determine a fundamental system of the homogeneous partial differential equation (6.4). Hence it is reasonable to work out a solution method of the Cauchy problem which does not need a fundamental system – that is, we do not need to know the "general solution". We can proceed as follows. Observe that if we apply the general theory of characteristics for the given quasilinear partial differential equation, then the auxiliary variable p does not occur in the characteristic equations for x and u:

$$x' = \partial_3 F(x, u, p) = f(x, u)\,,$$

$$u' = \langle \partial_F(x, u, p), p \rangle = f_0(x, u)\,.$$

This means that we need to solve this system only, with the following initial conditions:

$$x(t_0) = \varphi(s)\,,$$

$$u(t_0) = \psi(s)\,,$$

and then, using the solutions $t \mapsto X(t, s)$, $t \mapsto U(t, s)$, we express the variables t, s in terms of x form the system

$$x = X(t, s)$$

in the form $t = t(x)$, $s = s(x)$. Finally, the solution of the Cauchy problem is obtained by substitution into the equation $u = U(t, s)$:

$$u(x) = U\big(t(x), s(x)\big)\,.$$

As an example we solve the following Cauchy problem:

$$u\,\frac{\partial u}{\partial x} - \frac{\partial u}{\partial y} = 1\,,$$

$$u(-s, s) = 1\,.$$

Here we have $\varphi(s) = (-s, s)$, $\psi(s) = 1$. The equations of the characteristic system which refer to x and to u with the corresponding initial conditions are:

$$\begin{aligned} x' &= u & x(0) &= -s, \\ y' &= -1 & y(0) &= s, \\ u' &= 1 & u(0) &= 1. \end{aligned}$$

The solution of this system is

$$\begin{aligned} u(t) &= t + 1\,, \\ y(t) &= -t + s\,, \\ x(t) &= \frac{t^2}{2} + t - s\,. \end{aligned}$$

We express t and s from the last two equations in the form

$$t = \sqrt{2(x + y)}, \qquad s = y + \sqrt{2(x + y)}\,,$$

then we substitute them into the expression for u to obtain the solution of the Cauchy problem in the form

$$u(x, y) = \sqrt{2(x + y)} + 1\,.$$

7.6 Problems

Solve the following Cauchy problems:

1. $y^2 z_x + xy z_y = x, \quad z(0,y) = y^2$,
2. $xz_x - 2yz_y = x^2 + y^2, \quad z(x,1) = x^2$,
3. $$x\frac{\partial u}{\partial x} + y\frac{\partial u}{\partial y} = u - xy, \quad u(2,y) = y^2 + 1,$$
4. $$\tan x \frac{\partial u}{\partial x} + y\frac{\partial u}{\partial y} = u, \quad u(x,x) = x^3,$$
5. $$x\frac{\partial u}{\partial x} - y\frac{\partial u}{\partial y} = u^2(x - 3y), \quad x = 1,\ yu + 1 = 0,$$
6. $$x\frac{\partial u}{\partial x} + y\frac{\partial u}{\partial y} = u - x^2 - y^2, \quad u(x,-2) = x - x^2,$$
7. $$yu\frac{\partial u}{\partial x} + xu\frac{\partial u}{\partial y} = xy, \quad x = a,\ y^2 + u^2 = a^2.$$

7.7 Complete integral

For the sake of simplicity we shall study the concept of complete integral of first order partial differential equations in the two-variable case, only.

Let $\Omega \subseteq \mathbb{R}^5$, $D, \Gamma \subseteq \mathbb{R}^2$ be non-empty open sets, and let $F : \Omega \to \mathbb{R}$ be a continuous function. The function $\Phi : D \times \Gamma \to \mathbb{R}$ is called a *complete integral* of the first order partial differential equation

$$F(x, y, u, \partial_1 u, \partial_2 u) = 0 \tag{7.9}$$

on D, if Φ is twice continuously differentiable, the function

$$(x, y) \to \Phi(x, y, a, b)$$

is a solution of (7.9) on D for each (a, b) in Γ, and finally, the rank of the matrix

$$\frac{\partial(\Phi, \Phi_x, \Phi_y)}{\partial(a, b)}$$

is 2, whenever (a,b) is in Γ and (x,y) is in D. Roughly speaking, this last condition means that the parameters a,b are "independent".

Hence the complete integral is a two parameter family of solutions. We show how to find all solutions of (7.9), if a complete integral is available.

Let Φ be a complete integral of (7.9) on D. Then the system of partial differential equations

$$\Phi_a\big(x,y,a(x,y),b(x,y)\big)a_x(x,y)+\Phi_b\big(x,y,a(x,y),b(x,y)\big)b_x(x,y)=0\,,$$
$$\Phi_a\big(x,y,a(x,y),b(x,y)\big)a_y(x,y)+\Phi_b\big(x,y,a(x,y),b(x,y)\big)b_y(x,y)=0$$

for the unknown functions a,b is called the *Lagrange system* corresponding to the complete integral Φ. We shall use the Lagrange system for the following purpose: we try to determine the continuously differentiable functions $a,b:D\to\mathbb{R}$ such that the function $\varphi:D\to\mathbb{R}$ defined for each (x,y) in D by

$$\varphi(x,y)=\Phi\big(x,y,a(x,y),b(x,y)\big) \tag{7.10}$$

is a solution of (7.9) on D. As we have

$$\varphi_x=\Phi_x+\Phi_a\cdot a_x+\Phi_b\cdot b_x\,,$$
$$\varphi_y=\Phi_y+\Phi_a\cdot a_y+\Phi_b\cdot b_y\,,$$

it follows, by the Lagrange system, that $\varphi_x=\Phi_x$, and $\varphi_y=\Phi_y$, hence φ is a solution of (7.9) on D. Conversely, we show that, using an appropriate complete integral Φ, every solution φ of (7.9) can be obtained in the form (7.10).

Theorem 7.7.1. *Let $D,\Gamma\subseteq\mathbb{R}^2$ be non-empty open sets, let $\Phi:D\times\Gamma\to\mathbb{R}$ be a complete integral of (7.9) on D, further let (x_0,y_0) be in D and (a_0,b_0) in Γ such that*

$$\det\frac{\partial(\Phi,\Phi_x)}{\partial(a,b)}\bigg|_{(x_0,y_0,a_0,b_0)}\neq 0\,,$$

and

$$F(x_0,y_0,u_1,p_1,q_1)=F(x_0,y_0,u_2,p_2,q_2)$$

implies

$$u_1=u_2,p_1=p_2,q_1=q_2\,.$$

Then for each $\varphi:D\to\mathbb{R}$ twice continuously differentiable solution of (7.9) there exists a neighborhood $U\subseteq D$ of (x_0,y_0), and there exists a solution $a,b:U\to\mathbb{R}$ of the Lagrange system corresponding to Φ such that

$$\varphi(x,y)=\Phi\big(x,y,a(x,y),b(x,y)\big)$$

holds for each (x,y) in U.

Proof. We consider the implicit system of functions

$$\begin{aligned}\Phi(x,y,a,b)-\varphi(x,y)&=0\\ \Phi_x(x,y,a,b)-\varphi_x(x,y)&=0\\ \Phi_y(x,y,a,b)-\varphi_y(x,y)&=0\,,\end{aligned}$$

where φ is a solution of (7.9) on D. The Jacobian corresponding to the first two equations is non-zero, by the first condition of the theorem, and, by the second condition, the point (x_0,y_0,a_0,b_0) satisfies the first two equations. The functions $\mathcal{F},\mathcal{G}$, defined by

$$\begin{aligned}\mathcal{F}(x,y,a,b)&=\Phi(x,y,a,b)-\varphi(x,y)\,,\\ \mathcal{G}(x,y,a,b)&=\Phi_x(x,y,a,b)-\varphi_x(x,y)\,,\end{aligned}$$

are continuously differentiable on $D\times\Gamma$. Hence, by the Implicit Function Theorem, there exists a neighborhood $U\subseteq D$ of the point (x_0,y_0), and there are continuously differentiable functions $a,b:U\to\mathbb{R}$ such that

$$\Phi\big(x,y,a(x,y),b(x,y)\big)=\varphi(x,y)$$

holds for each (x,y) in U. □

Now we present some special types of partial differential equations for which a complete integral can be obtained easily. Let $\Gamma\subseteq\mathbb{R}^2$ be a non-empty open set, and let $f:\Gamma\to\mathbb{R}$ be a twice continuously differentiable function. By an easy computation we have that a complete integral of the *Clairut partial differential equation*

$$u=x\cdot\partial_x u+y\cdot\partial_y u+f(\partial_x u,\partial_y u) \tag{7.11}$$

is given by

$$\Phi(x,y,a,b)=ax+by+f(a,b)\,.$$

For instance, the Clairut differential equation

$$x\,u_x+y\,u_y+u_x\,u_y-u=0$$

has the following complete integral.

$$\Phi(x,y,a,b)=ax+by+ab\,.$$

Let $\Omega\subseteq\mathbb{R}^2$ be a non-empty open set, let $F:\Gamma\to\mathbb{R}$ be a twice continuously differentiable function, $I\subseteq\mathbb{R}$ a non-empty open interval, and $f:\Gamma\to\mathbb{R}$ a twice continuously differentiable function, further we assume that for each t in I the point $\big(t,f(t)\big)$ belongs to Ω, and $F\big(t,f(t)\big)=0$.

Then an easy computation shows that a complete integral of the partial differential equation

$$F(\partial_x u, \partial_y u) = 0 \tag{7.12}$$

is given by

$$\Phi(x, y, a, b) = ax + f(a)y + b\,.$$

We apply this for the partial differential equation

$$\frac{\partial u}{\partial x} = \left(\frac{\partial u}{\partial y}\right)^2.$$

to get the following complete integral:

$$\Phi(x, y, a, b) = ax \pm \sqrt{a}y + b\,.$$

Here obviously we have $f(t) = \pm\sqrt{t}$.

Let $\Omega \subseteq \mathbb{R}^3$ be a non-empty open set, and let $F : \Gamma \to \mathbb{R}$ be a twice continuously differentiable function. Let $I \subseteq \mathbb{R}$ be a non-empty open interval, and suppose that $\varphi : I \to \mathbb{R}$ is a non-constant twice continuously differentiable solution of the ordinary differential equation $F(\varphi, a\varphi', b\varphi') = 0$ on I, where a, b are real numbers. Then a simple calculation shows that a complete integral of the partial differential equation

$$F(u, \partial_x u, \partial_y u) = 0 \tag{7.13}$$

is given by

$$\Phi(x, y, a, b) = \varphi(ax + by)\,.$$

As an application, we consider the partial differential equation

$$u_x^2 + u_x u_y = u^2\,. \tag{7.14}$$

Here $F(u, p, q) = p^2 + pq - u^2$ and we look for a solution φ of the differential equation

$$a^2\varphi'^2 + ab\varphi'^2 - \varphi^2 = 0\,,$$

which also can be written in the form

$$\frac{\varphi'}{\varphi} = \frac{1}{a^2 + ab}\,.$$

Solving this first order ordinary differential equation we obtain the first integral

$$\Phi(x, y, a, b) = \varphi(ax + by) = \exp\frac{ax + by}{\sqrt{a^2 + ab}}\,.$$

Finally, we consider the special case of the *separable equation* which has the form

$$\varphi(x, u_x) = \psi(y, u_y)\,, \tag{7.15}$$

where φ, ψ are continuous on some open sets of $\mathbb{R}^2$. We introduce the auxiliary equation $\varphi(x, u_x) = a$, then also $\psi(y, u_y) = a$, where a is considered as a parameter. We write $\varphi(x, p) = a$ and $\psi(y, q) = a$ and we solve these equations in the form $p = f(x, a)$ and $q = g(y, a)$. Finally, by integrating the expression

$$f(x, a)dx + g(y, a)dy$$

as a total differential we obtain a complete integral in the form

$$\Phi(x, y, a, b) = \int f(x, a)dx + \int g(y, a)dy + b\,.$$

For instance, for the equation

$$u_x u_y = \frac{x}{y}$$

we have the complete integral

$$\Phi(x, y, a, b) = \int axdx + \int \frac{1}{ay}dy + b = a\frac{x^2}{2} + \frac{1}{a}\ln y + b\,.$$

Having a complete integral we can solve the Cauchy problem in the following way: using the initial curve $x = x(s)$, $y = y(s)$, $u = u(s)$ we consider the equation system

$$\begin{aligned} u(s) &= \Phi\big(x(s), y(s), a, b\big)\,, \\ u'(s) &= \Phi_x\big(x(s), y(s), a, b\big) \cdot x'(s) + \Phi_y\big(x(s), y(s), a, b\big) \cdot y'(s)\,, \end{aligned}$$

and we express a and b in terms of s in the form $a = a(s)$, $b = b(s)$. Finally, we determine the envelope of the one parameter family of surfaces

$$u = \Phi\big(x, y, a(s), b(s)\big)\,.$$

This can be proceeded by eliminating the parameter s from the system consisting of the equation

$$u = \Phi\big(x, y, a(s), b(s)\big)$$

and of its derivative with respect to s:

$$0 = \frac{d}{ds}\Big[\Phi\big(x, y, a(s), b(s)\big)\Big]\,.$$

As an example we solve the Cauchy problem

$$x\,u_x + y\,u_y + u_x\,u_y - u = 0\,,$$

$$u(0,y) = y^2$$

using complete integral.

The initial curve is

$$\begin{aligned} x(s) &= 0\,, \\ y(s) &= s\,, \\ u(s) &= s^2\,. \end{aligned}$$

By our above considerations, a complete integral is

$$\Phi(x,y,a,b) = ax + by + ab\,.$$

The system of equations for the functions a, b has the form

$$\begin{aligned} s^2 &= bs + ab\,, \\ 2s &= b\,. \end{aligned}$$

From this we derive $a(s) = -\frac{s}{2}$, $b(s) = 2s$, hence the one parameter family of surfaces is

$$u = -\frac{s}{2}x + 2sy - s^2\,.$$

The equation of the envelope can be obtained by eliminating s from the system

$$\begin{aligned} u &= -\frac{s}{2}x + 2sy - s^2\,, \\ 0 &= -\frac{x}{2} + 2y - 2s\,. \end{aligned}$$

From the second equation we have

$$s = y - \frac{x}{4}\,,$$

and substituting into the first equation it follows

$$\begin{aligned} u(x,y) &= -\frac{1}{2}\left(y - \frac{x}{4}\right)x + 2\left(y - \frac{x}{4}\right)y - \left(y - \frac{x}{4}\right)^2 \\ &= \frac{x^2}{16} - \frac{xy}{2} + y^2\,. \end{aligned}$$

7.8 Problems

1. Find a complete integral of the following partial differential equations:

   i) $u_x\, u_y - xy = 0\,,$
   ii) $u_x^2 + u_y^2 = 1\,,$
   iii) $u - u_x\, u_y = 0\,,$
   iv)
   $$\frac{u_x}{y} - \frac{u_y}{x} = \frac{1}{x} + \frac{1}{y},$$
   v) $u_x = \sin(x\, u_y)\,,$
   vi) $u_x^2 - u_y^3 = 0\,,$
   vii)
   $$u_x\, u_y = \frac{1}{2}.$$

2. Using complete integral find the solution of the given equations which fits to the given curve.

   i) $u_x^2 + u_y^2 = 1, \qquad x = \sin s, \quad y = \cos s, \quad u = 0\,,$
   ii) $x\, u_x + y\, u_y - u_x\, u_y = 0, \qquad u(0, y) = y\,.$

Chapter 8

HIGHER ORDER PARTIAL DIFFERENTIAL EQUATIONS

8.1 Special Cauchy problems for higher order partial differential equations

In order to formulate the special Cauchy problem for higher order partial differential equations we introduce new notation. Let m, n, N be positive integers, and let σ denote the set of those points $(k_0, k_1, \dots, k_n)$ in $\mathbb{N}^{n+1}$ for which $k_0 < m$ and $k_0 + k_1 + \cdots + k_n \leqslant N$ holds. The number of elements of σ will be denoted by $|\sigma|$. Let $\Omega \subseteq \mathbb{R} \times \mathbb{R}^n$ be a non-empty open set, further let $u : \Omega \to \mathbb{R}$ be a function having continuous partial derivatives $\partial_1^{k_0} \partial_2^{k_1} \dots \partial_{n+1}^{k_n} u$ for all those points $(k_0, k_1, \dots, k_n)$ which belong to σ. Obviously, the number of such partial derivatives is $|\sigma|$. Now, let $\Gamma \subseteq \mathbb{R} \times \mathbb{R}^n \times \mathbb{R}^{|\sigma|}$ and $D \subseteq \mathbb{R}^n$ be non-empty open sets, further let $F : \Gamma \to \mathbb{C}$ be a continuous function. We formulate a special Cauchy problem for the partial differential equation

$$\partial_1^m u + F\Big(t, x, u, \dots, \partial_1^{k_0} \partial_2^{k_1} \dots \partial_{n+1}^{k_n} u, \dots\Big) = 0 . \tag{8.1}$$

We can write (8.1) in the following, more suggestive form:

$$\partial_t^m u + F\Big(t, x, u, \dots, \frac{\partial^{k_0 + k_1 + \cdots + k_n} u}{\partial_t^{k_0} \partial_{x_1}^{k_1} \dots \partial_{x_n}^{k_n}}, \dots\Big) = 0 , \tag{8.2}$$

referring to the fact that we denote the elements of $\mathbb{R} \times \mathbb{R}^n$ by (t, x), where $x = (x_1, x_2, \dots, x_n)$. This means that the variable t plays a distinguished role, that's why we call this Cauchy problem "special". Another way of writing is the following:

$$\frac{\partial^m u}{\partial t^m} + F\Big(t, x, u, \dots, \frac{\partial^{k_0 + k_1 + \cdots + k_n} u}{\partial^{k_0} t\, \partial^{k_1} x_1 \dots \partial^{k_n} x_n}, \dots\Big) = 0 . \tag{8.3}$$

The function u is a *solution* of the partial differential equation (8.1) on Ω, if $\Omega \subseteq \mathbb{R} \times \mathbb{R}^n$ is a non-empty open set, and $u : \Omega \to \mathbb{C}$ is a function having all continuous partial derivatives which appear in, for each (t,x) in Ω the point

$$\big(t, x, u(t,x), \ldots, \partial_1^{k_0}\partial_2^{k_1}\ldots\partial_{n+1}^{k_n}u(t,x), \ldots\big)$$

belongs to Γ, further we have

$$\partial_1^m u(t,x) + F\big(t, x, u(t,x), \ldots, \partial_1^{k_0}\partial_2^{k_1}\ldots\partial_{n+1}^{k_n}u(t,x), \ldots\big) = 0\,.$$

Let $D \subseteq \mathbb{R}^n$ be a non-empty open set, t_0 a real number, further let $\varphi_i : D \to \mathbb{C}$ $(i = 0, 1, \ldots, m-1)$ be continuous functions. Then the system of equations

$$\partial_1^i u(t_0, x) = \varphi_i(x), \qquad (i = 0, 1, \ldots, m-1) \tag{8.4}$$

is called *special Cauchy problem* for the partial differential equation (8.1). The function u is a *solution* of the special Cauchy problem (8.4) on the set Ω, if for each element x in D the point (t_0, x) belongs to Ω, further the function $u : \Omega \to \mathbb{C}$ is a solution of the partial differential equation (8.1) on Ω such that for each x in D we have

$$\partial_1^i u(t_0, x) = \varphi_i(x), \qquad (i = 0, 1, \ldots, m-1)\,.$$

Sometimes we write equations (8.4) in the form

$$\partial_t^i u(t_0, x) = \varphi_i(x), \qquad (i = 0, 1, \ldots, m-1)\,,$$

respectively

$$\frac{\partial^i u}{\partial t^i}(t_0, x) = \varphi_i(x), \qquad (i = 0, 1, \ldots, m-1)\,,$$

indicating the distinguished role of the variable t. The following special cases are particularly important in physics: in the case $m = 1$, $N = 2$ we have

$$\begin{aligned} \partial_t u &= \sum_{i=1}^{n} \partial_i^2 u + f(t,x)\,, \\ u(t_0, x) &= \varphi(x)\,, \end{aligned}$$

which is called the *special Cauchy problem for the n-dimensional heat equation*, and in the case $m = 2$, $N = 2$ we have

$$\begin{aligned} \partial_t^2 u &= \sum_{i=1}^{n} \partial_i^2 u + f(t,x)\,, \\ u(t_0, x) &= \varphi(x)\,, \\ \partial_t u(t_0, x) &= \psi(x)\,, \end{aligned}$$

which is called the *special Cauchy problem for the n-dimensional wave equation.* We get another special case if F in (8.1) depends linearly on the partial derivatives. More exactly, let $\Omega \subseteq \mathbb{R}^{\times}\mathbb{R}^n$ be a non-empty open set, for each $(k_0, k_1, \ldots, k_n)$ in σ let $A_{k_0,k_1,\ldots,k_n} : \Omega \to \mathbb{C}$ be a continuous function, further let $f : \Omega \to \mathbb{C}$ be a continuous function, too. Then the partial differential equation

$$\partial_1^m u + \sum_{(k_0,k_1,\ldots,k_n)\in\sigma} A_{k_0,k_1,\ldots,k_n}(t,x)\, \partial_1^{k_0} \partial_2^{k_1} \ldots \partial_{n+1}^{k_n} u = f(t,x) \tag{8.5}$$

is called *linear.* If, in addition, the coefficient functions $A_{k_0,k_1,\ldots,k_n}$ are constant on Ω, then the equation is called *linear equation with constant coefficients.* The heat equation and the wave equation are important examples for this type.

We note that if all the given functions in the above problems are real valued then it makes sense to look for real valued solutions which are simply called *real solutions.*

8.2 Theorems of Kovalevskaya and Holmgren

The following two theorems on special Cauchy problems for higher order partial differential equations are of fundamental importance which we present here without proof. We use the notation of the preceding section.

Theorem 8.2.1. *(Kovalevskaya) Let the F, φ_i $(i = 0, 1, \ldots, m-1)$ in the special Cauchy problem* (8.1), (8.4) *be analytic functions. Then every point (t_0, x_0) in Ω has a neighborhood in which the special Cauchy problem* (8.1), (8.4) *has a unique analytic solution.*

We note that the special Cauchy problem (8.1), (8.4) can be generalized, and under some conditions the more general problem can be reduced to the special Cauchy problem. This is the basic importance of this existence and uniqueness theorem. S. Kovalevskaya gave an example for a problem of the above type with non-analytic given functions, and the problem has no analytic solution. Her example is the following:

$$\begin{aligned} \partial_t u &= \partial_x^2 u \\ u(0,x) &= \frac{1}{1+x^2}\,, \end{aligned}$$

which has a unique formal power series solution around $(0,0)$. The following natural question arises: if the given functions are real valued and continuously differentiable as many times as it is necessary for the problem makes sense, then does the problem has a solution? So far this question has remained open.

For the special Cauchy problem we have the following uniqueness result.

Theorem 8.2.2. *(Holmgren) In the special Cauchy problem* (8.5), (8.4) *let the functions* $A_{k_0,k_1,\ldots,k_n}$ *be analytic for each* $(k_0, k_1, \ldots, k_n)$ *in* σ. *Then every point* (t_0, x_0) *in* Ω *has a neighborhood in which the problem* (8.5), (8.4) *has at most one solution.*

This uniqueness result has been proved by H. J. Holmgren in 1901. Later J. Hadamard proved that the study of uniqueness of the real solution for linear problems can be reduced to the study of uniqueness of linear problems with sufficiently smooth (but not necessarily analytic) coefficients. Then the investigations focused on this latter problem. Finally, this problem has been solved by T. Carleman in 1938 in the case of two independent variables with twice continuously differentiable coefficient functions.

8.3 Linear partial differential operators

In this section we study an important class of operators, the so-called *linear partial differential operators.* We shall use multi-index notation (see Section 5.21).

The partial differential operators ∂_j defined on appropriate classes of functions have the usual meaning. We use the vector differential operator $\partial = (\partial_1, \partial_2, \ldots, \partial_n)$ and we compute with this vector formally in the same manner as we do it with proper vectors. Hence, for every multi-index α we write

$$\partial^\alpha = \partial_1^{\alpha_1} \cdot \partial_2^{\alpha_2} \ldots \partial_n^{\alpha_n} .$$

We note that ∂_j^0 always means the identity operator which leaves every function fixed.

Let $\Omega \subseteq \mathbb{R}^n$ be an open set, let N be a natural number, and for each multi-index α with $|\alpha| \leqslant N$ let $c_\alpha : \Omega \to \mathbb{C}$ be a continuous function. We define the differential operator

$$P(x, \partial) = \sum_{|\alpha| \leqslant N} c_\alpha(x) \partial^\alpha \tag{8.6}$$

for x in Ω. Then for each N-times continuously differentiable function $f : \Omega \to \mathbb{C}$ we define

$$P(x,\partial)f(x) = \sum_{|\alpha| \leqslant N} c_\alpha(x)\partial_1^{\alpha_1}\partial_2^{\alpha_2}\dots\partial_n^{\alpha_n} f(x)$$

for x in Ω. The operator $P(x,\partial)$ is called a *linear partial differential operator*. The number N is called the *order* of $P(x,\partial)$ if there exists an α with $|\alpha| = N$ such that c_α is not identically zero on Ω. The functions c_α are called the *coefficients* of $P(x,\partial)$. The *principal part* of $P(x,\partial)$ is the operator

$$P_f(x,\partial) = \sum_{|\alpha| = N} c_\alpha(x)\partial^\alpha .$$

If c_α is constant on Ω for each α with $|\alpha| \leqslant N$, then we say that $P(x,\partial)$ is an operator with *constant coefficients*. In this case the polynomial

$$P(\lambda) = \sum_{|\alpha| \leqslant N} c_\alpha \lambda^\alpha \tag{8.7}$$

is called the *characteristic polynomial* of the operator $P(\partial) = P(x,\partial)$. The operator $P(\partial)$ is called *homogeneous*, if the characteristic polynomial P is a homogeneous polynomial. Obviously, the principal part of each linear partial differential operator with constant coefficients is a homogeneous operator.

Here we recall some classical basic examples.

1) **The Laplace operator** The *Laplace operator*, or *Laplacian* is the following differential operator:

$$P(\partial) = \sum_{j=1}^{n} \partial_j^2 . \tag{8.8}$$

It is a second order homogeneous linear partial differential operator with constant coefficients. Its characteristic polynomial is

$$P(\lambda) = \sum_{j=1}^{n} \lambda_j^2 = \|\lambda\|^2 .$$

The Laplace operator is denoted by Δ. The partial differential equation

$$\Delta u = 0 \tag{8.9}$$

is called *Laplace equation*, and its solutions on any open set $\Omega \subseteq \mathbb{R}^n$ are called *harmonic functions* on Ω.

The Laplace operator has the following crucial property.

Theorem 8.3.1. *Let $T : \mathbb{R}^n \to \mathbb{R}^n$ be an orthogonal linear transformation and $f : \mathbb{R}^n \to \mathbb{R}$ an infinitely differentiable function. Then we have*

$$(\Delta f) \circ T = \Delta(f \circ T)\,. \tag{8.10}$$

This property of the Laplace operator can be expressed by saying that the Laplacian is invariant under orthogonal transformations.

2) **The wave operator** We use the following notation: the coordinates of a point in $\mathbb{R}\times\mathbb{R}^n$ will be written in the form $(t, x_1, x_2, \ldots, x_n)$, and the corresponding partial differential operators we denote by $\partial_t, \partial_1, \partial_2, \ldots, \partial_n$, respectively. We denote by Δ_x the Laplacian with respect to the x-components, that is

$$\Delta_x = \sum_{j=1}^{n} \partial_j^2\,.$$

Using this notation the *wave operator* is the following differential operator:

$$P(\partial) = \partial_t^2 - \Delta_x = \partial_t^2 - \sum_{j=1}^{n} \partial_j^2\,. \tag{8.11}$$

It is a second order homogeneous linear partial differential operator with constant coefficients. Its characteristic polynomial is

$$P(\mu, \lambda) = \mu^2 - \sum_{j=1}^{n} \lambda_j^2\,, \tag{8.12}$$

where $P : \mathbb{R} \times \mathbb{R}^n \to \mathbb{R}$, and the points in $\mathbb{R} \times \mathbb{R}^n$ are denoted by $(\mu, \lambda) = (\mu, \lambda_1, \lambda_2, \ldots, \lambda_n)$. The wave operator is denoted by $\Box$. The partial differential equation

$$\Box u = 0 \tag{8.13}$$

is called *wave equation.* If we are interested in those linear transformations in $\mathbb{R}\times\mathbb{R}^n$ which commute with $\Box$, then we find easily that they are the same as the linear transformations which leave the quadratic form (8.12) invariant. They form the so-called *Lorentz group.*

3) **The heat operator** We use the notation of the previous example. The *heat operator* is the following differential operator:

$$P(\partial) = \partial_t - \Delta_x = \partial_t^2 - \sum_{j=1}^{n} \partial_j^2\,. \tag{8.14}$$

It is a second order linear partial differential operator with constant coefficients. Its characteristic polynomial is

$$P(\mu, \lambda) = \mu - \sum_{j=1}^{n} \lambda_j^2 . \tag{8.15}$$

The partial differential equation

$$\partial_t u = \Delta_x u \tag{8.16}$$

is called *heat equation.*

4) **The Cauchy–Riemann operator** We denote an arbitrary point in $\mathbb{R}^2$ by (x, y) further let $\partial_x = \partial_1$, and $\partial_y = \partial_2$. We introduce the *Cauchy–Riemann operator* in the following way:

$$P(\partial) = \frac{1}{2}(\partial_x + i\,\partial_y)\,, \tag{8.17}$$

which is a first order linear homogeneous partial differential operator with constant coefficients. Its characteristic polynomial is

$$P(\lambda, \mu) = \frac{1}{2}(\lambda + i\mu)\,, \tag{8.18}$$

where (λ, μ) is arbitrary in $\mathbb{R}^2$. The corresponding partial differential equation

$$\partial_x f + i\partial_y f = 0 \tag{8.19}$$

is called *Cauchy–Riemann equation*, and here $f : \mathbb{R}^2 \to \mathbb{C}$ is a complex valued function of the form $f = u+iv$, and u, v are real valued. Equation (8.19) is equivalent to the following system of equations:

$$\partial_x u = \partial_y v, \quad \partial_y u = -\partial_x v\,. \tag{8.20}$$

Introducing the new variables $z = x + iy$ and $\bar{z} = x - iy$ in $\mathbb{R}^2$ we have $x = \frac{1}{2}(z + \bar{z})$, $y = \frac{1}{2i}(z - \bar{z})$, and (8.19) can be written in the form

$$\partial_{\bar{z}} f = 0\,. \tag{8.21}$$

The Cauchy–Riemann operator can be denoted by $\partial_{\bar{z}}$. Roughly speaking, the above equation says that "f is independent of $\bar{z}$". In complex function theory the solutions of the Cauchy–Riemann equation are called *holomorphic functions.* It is reasonable to introduce also the "anti-Cauchy–Riemann" operator by

$$\partial_z = \frac{1}{2}(\partial_x - i\,\partial_y)\,. \tag{8.22}$$

Then we have $\partial_{\bar{z}}\partial_z = \Delta$, the Laplacian in two variables.

8.4 Problems

1. Prove Theorem 8.3.1.
2. Prove the converse of Theorem 8.3.1: if T is a linear transformation in $\mathbb{R}^n$ satisfying (8.10) for every infinitely differentiable function f, then T is an orthogonal tranformation.
3. Let $\square = \partial_t^2 - \partial_x^2$ be the wave operator in $\mathbb{R} \times \mathbb{R}$. Show that there is a linear transformation in $\mathbb{R}^2$ transforming $\square$ into $\partial_1 \partial_2$.
4. Show that the twice continuously differentiable solutions of the wave equation in $\mathbb{R}^2$ have the form

$$u(t, x) = f(t + x) + g(t - x)\,,$$

where $f, g : \mathbb{R} \to \mathbb{R}$ are arbitrary twice continuously differentiable functions.

8.5 Exponential polynomial solutions of partial differential equations

In Section 5.21 we have seen that the Fourier transform of exponential polynomials can be applied to find exponential polynomial solutions of ordinary differential equations. It turns out that the basic formula in Theorem 5.21.4 can be used to find exponential polynomial solutions of inhomogeneous linear partial differential equations with constant or polynomial coefficients, if the inhomogeneous term is an exponential polynomial. Now we illustrate this on the Cauchy problem for the heat equation, that is on the problem

$$\begin{aligned} \partial_t u &= a^2 \Delta_x u\,, \\ u(0, x) &= u_0(x), \end{aligned} \tag{8.23}$$

where $u_0 : \mathbb{R}^n \to \mathbb{C}$ is a given non-zero exponential polynomial and $a > 0$ is a real number. We use the notation of Section 8.3 and we try to find an exponential polynomial $u : \mathbb{R} \times \mathbb{R}^n \to \mathbb{C}$, which is a solution of (8.23). We note that from the general theory it follows that the solution is unique, also in the case of the other equations in this section.

Observing that all exponentials m on $\mathbb{R} \times \mathbb{R}^n$ have the form

$$m(x, t) = \exp(\mu t + \langle \lambda, x \rangle)$$

with some μ in $\mathbb{C}$ and λ in $\mathbb{C}^n$, the value of the Fourier transform of u at the exponential m will be denoted by $\widehat{u}(\mu, \lambda)$. Applying Fourier transformation on both sides of the first equation of (8.23) we have, by the formula in Theorem 5.21.4

$$\partial_t \widehat{u}(\mu, \lambda) + \mu \cdot \widehat{u}(\mu, \lambda) = a^2 (\Delta_x + 2\langle \lambda, \partial \rangle + \langle \lambda, \lambda \rangle)\, \widehat{u}(\mu, \lambda) . \tag{8.24}$$

Here $\langle \lambda, \partial \rangle = \sum_{i=1}^n \lambda_i \partial_i$. We know that $\widehat{u}(\mu, \lambda)$ is a polynomial in (x, t) for every μ in $\mathbb{C}$, and for all λ in $\mathbb{C}^n$. For a fixed pair μ, λ we let

$$\widehat{u}(\mu, \lambda)(t, x) = a_N(x) t^N + a_{N-1}(x) t^{N-1} + \cdots + a_0(x) ,$$

where a_k is a polynomial for $k = 0, 1, \ldots, N$ and $a_N \neq 0$. Substituting into (8.24) and comparing the coefficients of t^N we have $\mu = a^2 \langle \lambda, \lambda \rangle$. Then we continue with comparing the coefficients of t^k for $k = 0, 1, \ldots, N-1$, and we have

$$a_{k+1}(x) = \frac{a^2}{k+1} (\Delta_x + 2\langle \lambda, \partial \rangle)\, a_k(x)$$

for all x in $\mathbb{R}^n$. Obviously, $a_0(x) = \widehat{u}_0(\lambda)(x)$, hence

$$a_k(x) = \frac{a^2}{k!} (\Delta_x + 2\langle \lambda, \partial \rangle)^k \widehat{u}_0(\lambda)(x)$$

holds for $k = 0, 1, \ldots, N$. Here N denotes the smallest non-negative integer for which $a_N \neq 0$ and $a_{N+1} = 0$. The existence of such N follows from the fact that $\widehat{u}_0(\lambda)$ is a polynomial, which is non-zero.

Using the inversion formula in Theorem 5.21.3 we have

Theorem 8.5.1. *Let $u_0 : \mathbb{R}^n \to \mathbb{C}$ be a non-zero exponential polynomial. Then the unique solution of the Cauchy problem (8.23) is given by*

$$u(t, x) = \sum_{\lambda \in \mathbb{C}^n} \sum_{N=0}^{\infty} \frac{[a^2(\Delta_x + 2\langle \lambda, \partial \rangle)]^N}{N!} \widehat{u}_0(\lambda)(x) \cdot t^N \cdot e^{a^2 \langle \lambda, \lambda \rangle t + \langle \lambda, x \rangle}$$

for all t in $\mathbb{R}$ and x in $\mathbb{R}^n$.

Here both sums are actually finite. A straightforward extension is the corresponding result for the problem

$$\partial_t u = a^2 \Delta_x u , \tag{8.25}$$
$$u(t_0, x) = u_0(x) ,$$

where $u_0 : \mathbb{R}^n \to \mathbb{C}$ is an exponential polynomial, and t_0 is a real number.

Theorem 8.5.2. *Let $u_0 : \mathbb{R}^n \to \mathbb{C}$ be a non-zero exponential polynomial and t_0 a real number. Then the unique solution of the Cauchy problem (8.25) is given by*

$$u(t, x) = \sum_{\lambda \in \mathbb{C}^n} \sum_{N=0}^{\infty} \frac{[a^2(\Delta_x + 2\langle \lambda, \partial \rangle)]^N}{N!} \widehat{u}_0(\lambda)(x) \cdot (t - t_0)^N \cdot e^{a^2 \langle \lambda, \lambda \rangle (t - t_0) + \langle \lambda, x \rangle} \tag{8.26}$$

for all t in $\mathbb{R}$ and x in $\mathbb{R}^n$.

The next step is to solve the inhomogeneous Cauchy problem

$$\begin{aligned} \partial_t u &= a^2 \Delta_x u + f(t,x)\,, \\ u(0,x) &= u_0(x)\,. \end{aligned} \tag{8.27}$$

Here we suppose that $u_0 : \mathbb{R}^n \to \mathbb{C}$ and $f : \mathbb{R} \times \mathbb{R}^n \to \mathbb{C}$ are exponential polynomials and we are looking for an exponential polynomial solution u of the problem (8.27). Our basic observation is that if the function $(t,x) \mapsto v(t,x,t_0)$ is a solution of the problem (8.25) with $u_0(x) = f(t_0,x)$, then the function

$$(t,x) \mapsto \int_0^t v(t,x,\tau)\,d\tau$$

is a solution of

$$\begin{aligned} \partial_t u &= a^2 \Delta_x u + f(t,x)\,, \\ u(0,x) &= 0\,. \end{aligned} \tag{8.28}$$

This can be verified by direct computation, and it makes possible to reduce the inhomogeneous problem (8.27) to the homogeneous one (8.28).

Using the formula (8.26) we infer that the function

$$u(t,x) = \sum_{\mu \in \mathbb{C}} \sum_{\lambda \in \mathbb{C}^n} \sum_{N=0}^{\infty} \int_0^t \cdots \tag{8.29}$$

$$\frac{[a^2(\Delta_x + 2\langle \lambda, \partial \rangle)]^N}{N!} \hat{f}(\mu,\lambda)(\tau,x) \cdot (t-\tau)^N \cdot e^{\mu\tau + a^2\langle \lambda,\lambda\rangle(t-\tau) + \langle \lambda, x\rangle}\,d\tau$$

is a solution of (8.28). Using Taylor's Formula we obtain

$$\hat{f}(\mu,\lambda)(\tau,x) = \sum_{k=0}^{\infty} \frac{\partial_{n+1}^k \hat{f}(\mu,\lambda)(t,x)}{k!}(-1)^k(t-\tau)^k\,,$$

which is a finite sum as $\hat{f}(\mu,\lambda)$ is a polynomial. From this equation we can see that the only job is to compute integrals of the form

$$\int_0^t (t-\tau)^{N+k} e^{[a^2\langle \lambda,\lambda\rangle - \mu](t-\tau)}\,d\tau\,.$$

Now we put

$$I_M(\alpha) = \int_0^t s^M e^{\alpha s}\,ds$$

for each non-negative integer M and complex number α. Then the following theorem is an easy consequence of Theorems 8.5.1 and 8.5.2.

Theorem 8.5.3. *Let $u_0 : \mathbb{R}^n \to \mathbb{C}$ and $f : \mathbb{R} \times \mathbb{R}^n \to \mathbb{C}$ be exponential polynomials and let $a > 0$ be a real number. Then the unique solution u of the Cauchy problem (8.27) can be written in the form $u = u_1 + u_2$, where*

$$u_1(t,x) = \sum_{\lambda\in\mathbb{C}^n} \sum_{N=0}^{\infty} \frac{[a^2(\Delta_x + 2\langle\lambda,\partial\rangle)]^N}{N!} \widehat{u}_0(\lambda)(x) \cdot t^N \cdot e^{a^2\langle\lambda,\lambda\rangle t + \langle\lambda,x\rangle}$$

and

$$u_2(t,x) = \sum_{\mu\in\mathbb{C}} \sum_{\lambda\in\mathbb{C}^n} \sum_{N=0}^{\infty} \sum_{k=0}^{\infty} \dots$$

$$(-1)^k \frac{[a^2(\Delta_x + 2\langle\lambda,\partial\rangle)]^N}{N!k!} \partial_{n+1}^k \widehat{f}(\mu,\lambda)(t,x) \cdot I_{N+k}(a^2\langle\lambda,\lambda\rangle - \mu) \cdot e^{\mu t + \langle\lambda,x\rangle}$$

for each t in $\mathbb{R}$ and x in $\mathbb{R}^n$. Here we use the notation

$$I_M(\alpha) = e^{\alpha t} \sum_{j=0}^{M} (-1)^j \frac{M!}{(M-j)!} \frac{t^{M-j}}{\alpha^{j+1}} - (-1)^M \frac{M!}{\alpha^{M+1}}, \quad \text{and } I_M(0) = \frac{t^{M+1}}{M+1}$$

for $\alpha \neq 0$.

These ideas extend easily also to Cauchy problems for other types of linear partial differential equations, if the given functions are exponential polynomials. A general type of these equations are those of the *evolution type equations*:

$$\begin{aligned} \partial_t u &= P(\partial)u + f(t,x) \\ u(0,x,y) &= u_0(x)\,. \end{aligned} \tag{8.30}$$

Here P is a complex polynomial in n variables, and f, u_0 are exponential polynomials. By the above method we can find all exponential polynomial solutions, and if uniqueness is guaranteed then we determine the only solution. For such problems we have the following theorem, which can be verified by direct computation.

Theorem 8.5.4. *Let $u_0 : \mathbb{R}^n \to \mathbb{C}$ and $f : \mathbb{R} \times \mathbb{R}^n \to \mathbb{C}$ be exponential polynomials and let $P : \mathbb{R}^n \to \mathbb{C}$ be a polynomial. Then the exponential polynomial $u = u_1 + u_2$ is a solution of the Cauchy problem (8.30), where*

$$u_1(t,x) = \sum_{\lambda\in\mathbb{C}^n} \sum_{N=0}^{\infty} \frac{[P(\partial+\lambda) - P(\lambda)]^N}{N!} \widehat{u}_0(\lambda)(x) \cdot t^N \cdot e^{P(\lambda)t + \langle\lambda,x\rangle}$$

and

$$u_2(t,x) = \sum_{\mu\in\mathbb{C}} \sum_{\lambda\in\mathbb{C}^n} \sum_{N=0}^{\infty} \sum_{k=0}^{\infty}$$

$$(-1)^k \frac{[P(\partial+\lambda)-P(\lambda)]^N}{N!k!}\, \partial_{n+1}^k \widehat{f}(\mu,\lambda)(t,x)\cdot I_{N+k}(P(\lambda)-\mu)\cdot e^{\mu t+\langle\lambda,x\rangle}$$

for each t *in* $\mathbb{R}$ *and* x *in* $\mathbb{R}^n$.

As an application we solve the Cauchy problem for the Schrödinger equation

$$\begin{aligned} \partial_t u &= i\Delta u + x\cos t - y^2 \sin t \\ u(0,x,y) &= x^2 + y^2 . \end{aligned}$$

Here $\Delta = \partial_1^2 + \partial_2^2$ is the Laplacian in $\mathbb{R}^2$. Using the notation of Theorem 8.5.4 we have $P(\xi,\eta) = i(\xi^2+\eta^2)$, further $u_0(x,y) = x^2+y^2$, and $f(t,x,y) = x\cos t - y^2 \sin t$. It follows

$$\widehat{u}_0(\lambda_1,\lambda_2)(x,y) = \begin{cases} x^2+y^2 & \text{for } \lambda_1=\lambda_2=0 \\ 0 & \text{otherwise}, \end{cases}$$

$$\widehat{f}(\mu,\lambda_1,\lambda_2)(t,x,y) = \begin{cases} \frac{x}{2} - \frac{y^2}{2i} & \text{for } \mu = i,\, \lambda_1=\lambda_2=0 \\ \frac{x}{2} + \frac{y^2}{2i} & \text{for } \mu = -i,\, \lambda_1=\lambda_2=0 \\ 0 & \text{otherwise}, \end{cases}$$

and finally

$$I_0(\pm i) = \frac{e^{\pm i}-1}{\pm i}, \quad I_1(\pm i) = \frac{te^{\pm it}}{\pm i} + e^{\pm it} - 1 ,$$

hence, by the previous theorem, we have that

$$\begin{aligned} u_1(t,x,y) &= x^2 + y^2 + 4it \\ u_2(t,x,y) &= x\sin t + y^2(\cos t - 1) - 2it + 2i\sin t , \end{aligned}$$

and the solution is

$$u(t,x,y) = u_1(t,x,y) + u_2(t,x,y) = x\sin t + x^2 + y^2\cos t + 2i(t+\sin t) .$$

8.6 Problems

1. Solve the following problems.

   i) $u_t = 4u_{xx} + t + e^t, \quad u(0,x) = 2$
   ii) $u_t = u_{xx} + 3t^2, \quad u(0,x) = \sin x$
   iii) $u_t = u_{xx} + e^{-t}\cos x, \quad u(0,x) = \cos x$
   iv) $u_t = u_{xx} + e^t \sin x, \quad u(0,x) = \sin x$

2. Solve the following problems.

   i) $u_t = \Delta_{xy} u + e^t, \quad u(0,x,y) = \cos x \sin y$
   ii) $u_t = \Delta_{xy} u + \sin t \sin x \sin y \quad u(0,x,y) = 1$

3. Solve the following problems.

   i) $u_t = 2\Delta_{xyz} u + t\cos x, \quad u(0,x,y,z) = \cos y \cos z$
   ii) $u_t = 3\Delta_{xyz} u + e^t \quad u(0,x,y,z) = \sin(x - y - z)$

4. Solve the following problems.

   i) $u_t = iu_{xx} + tx^3, \quad u(0,x) = x^4$
   ii) $u_t = i\Delta_{xy} u + x\cos t - y^2 \sin t \quad u(0,x,y) = x^2 + y^2$
   iii) $u_t = i\Delta_{xyz} u + 6x + y^2 + iz^3, \quad u(0,x,y,z) = i(x^3 + y^3 + z^3)$

Chapter 9

SECOND ORDER QUASILINEAR PARTIAL DIFFERENTIAL EQUATIONS

9.1 Second order partial differential equations with linear principal part

In this section n denotes a positive integer. Let $\Omega \subseteq \mathbb{R}^n \times \mathbb{R} \times \mathbb{R}^n$, $D \subseteq \mathbb{R}^n$ non-empty open sets, $f : \Omega \to \mathbb{R}$ a continuous function, and let $a_{i,k} : D \to \mathbb{R}$ $(i, k = 1, 2, \ldots, n)$ continuous functions such that $a_{i,k} = a_{k,i}$ holds for each $i, k = 1, 2, \ldots, n$. For each x in D let $A(x)$ denote the $n \times n$ matrix with entries $a_{i,k}(x)$. Let $\mathcal{C}^2(D, \Omega)$ denote the set of those twice continuously differentiable functions $u : D \to \mathbb{R}$ with the property that for each x in D the point $\big(x, u(x), u'(x)\big)$ lies in Ω, further let $\mathcal{C}(D)$ denote the set of all continuous real valued functions on D. If u is in $\mathcal{C}^2(D, \Omega)$, and x is a point in D, then we let

$$Lu(x) = \sum_{i=1}^{n} \sum_{k=1}^{n} a_{i,k}(x)\, \partial_i \partial_k\, u(x) + f\big(x, u(x), u'(x)\big)\,.$$

The differential operator $L : \mathcal{C}^2(D, \Omega) \to \mathcal{C}(D)$ is called *second order differential operator with linear principal part* on D. The matrix $A(x)$ is called the *matrix of the operator* L at the point x, and the expression

$$\sum_{i=1}^{n} \sum_{k=1}^{n} a_{i,k}(x)\, \partial_i \partial_k\, u(x)$$

is called the *principal part of* L. Using the trace of linear operators we can write the L in the compact form

$$Lu(x) = \mathrm{Tr}\,\big(A(x) u''(x)\big) + f\big(x, u(x), u'(x)\big)\,.$$

Let x_0 be a point in D. The operator L is called *elliptic*, *parabolic*, or *hyperbolic* in the point x_0, if the matrix $A(x_0)$ is definite, semidefinite, or indefinite, respectively. We say that L is *elliptic, parabolic,* or *hyperbolic* on a non-empty, open set $D_0 \subseteq D$, if it is of the given type at each point of D_0. We say that L has *normal form* at the point x_0, if $A(x_0)$ is a diagonal matrix, and we say that it has normal form on D_0, if it has normal form at each point of D_0.

The equation

$$Lu = 0 \tag{9.1}$$

is called *second order partial differential equations with linear principal part.*

9.2 Problems

1. Find all solutions of the given partial differential equations on $\mathbb{R}^2$ ($f : \mathbb{R}^2 \to \mathbb{R}$ is a given continuous function and $a > 0$ is a constant).

   i)
$$\frac{\partial^2 u}{\partial x\, \partial y} = f(x, y),$$

   ii)
$$\frac{\partial^2 u}{\partial x\, \partial y} - \frac{\partial u}{\partial x} = f(x, y),$$

   iii)
$$\frac{\partial^2 u}{\partial x^2} - \frac{\partial^2 u}{\partial y^2} = 0,$$

   iv)
$$\frac{\partial^2 u}{\partial x^2} - a^2 \frac{\partial^2 u}{\partial y^2} = 0,$$

   v)
$$\frac{\partial^2 u}{\partial x^2} = \frac{\partial^2 u}{\partial x\, \partial y}.$$

2. Find all solutions of the given partial differential equations on $\mathbb{R}^2$ satisfying the given side conditions.

   i)
$$\frac{\partial^2 u}{\partial x\, \partial y} = x + y, \qquad u(x, x) = x, \qquad u_x(x, x) = 0,$$

ii)

$$\frac{\partial^2 u}{\partial x^2} = \frac{\partial^2 u}{\partial y^2}, \qquad u(0,y) = 1, \qquad u_x(0,y) = 1.$$

3. Find all solutions of the given partial differential equation of the given special form.

   i)

$$\frac{\partial^2 u}{\partial x^2} - \frac{\partial u}{\partial y} = 0, \qquad u(x,y) = A(x) \cdot B(y),$$

   ii)

$$\frac{\partial^2 u}{\partial x^2} + \frac{\partial^2 u}{\partial y^2} = 1, \qquad u(x,y) = A(x) + B(y),$$

   iii)

$$\frac{\partial^2 u}{\partial x^2} + \frac{\partial^2 u}{\partial y^2} + \frac{\partial^2 u}{\partial z^2} = 1, \qquad u(x,y,z) = A(x) + B(y) + C(z).$$

9.3 Linear transformation, normal form

We use the notation of the previous section. Let x_0 be a point in D, and let $S : \mathbb{R}^n \to \mathbb{R}^n$ be an orthogonal linear transformation such that the matrix

$$B(x) = S^{-1}\, A(x)\, S$$

is diagonal at x_0. The orthogonality of S means that it is regular, and its inverse is equal to its transpose: $S^{-1} = S^t$. If y is arbitrary in $S^{-1}(D)$, then we let

$$v(y) = u(S\, y)\,.$$

Then for each x in D we have $v\big(S^t(x)\big) = u(x)$, further

$$\begin{aligned} u'(x) &= v'(S^t x) \cdot S^t\,, \\ u'^{\,t}(x) &= S \cdot v'^{\,t}(S^t x)\,, \\ u''^{\,t}(x) &= S \cdot v''^{\,t}(S^t x) \cdot S^t\,. \end{aligned}$$

As u'' and v'' are symmetric, hence we have

$$u''(x) = S \cdot v''(S^t x) \cdot S^t$$

and

$$Lu(x) = \operatorname{Tr}\big(A(x) \cdot u''(x)\big) + f\big(x, u(x), u'(x)\big)$$

$$
\begin{aligned}
&= \operatorname{Tr}\left(A(x)\cdot S\cdot v''(S^t x)\cdot S^t\right) + f\big(x,u(x),u'(x)\big)\\
&= \operatorname{Tr}\left(S^t\cdot A(x)\cdot S\cdot v''(S^t x)\right) + f\big(x,u(x),u'(x)\big)\\
&= \operatorname{Tr}\left(B(x)\cdot v''(S^t x)\right) + f\big(x,v(S^t x),v'(S^t x)S^t\big)\\
&= \operatorname{Tr}\left(C(y)\cdot v''(y)\right) + g\big(y,v(y),v'(y)\big)\\
&= Kv(y)\,,
\end{aligned}
$$

where

$$
\begin{aligned}
y &= S^t x = S^{-1}x\,,\\
C(y) &= B(S\,y)\,,\\
g(y,v,p) &= f(S\,y,u,S^t\,p)\,,
\end{aligned}
$$

and K is of the same type as L on $S^{-1}(D)$. As $B(x_0)$ is a diagonal matrix, so is $C(y_0)$, where $y_0 = S^t x_0$, hence the operator K has normal form at the point y_0.

It can be shown that, in general, it is not possible to transform every second order partial differential operator with linear principal part to normal form at every point of an open set, if the number of variables is more than 2 – even when applying nonlinear transformations.

9.4 Problems

1. Find a normal form for the following partial differential equations:
   - i) $u_{xx} + 2u_{xy} - 2u_{xz} + 2u_{yy} + 6u_{zz} = 0$
   - ii) $4u_{xx} - 4u_{xy} - 2u_{yz} + u_y + u_z = 0$
   - iii) $u_{xy} - u_{xz} + u_y + u_y - u_z = 0$
   - iv) $u_{xx} + 2u_{xy} - 4u_{xz} - 6u_{yz} - u_{zz} = 0$
   - v) $u_{xx} + 2u_{xy} + 2u_{yy} + 2u_{yz} + 2u_{yt} + 2u_{zz} + 3u_{tt} = 0$
   - vi) $u_{xy} - u_{xt} + u_{zz} - 2u_{zt} + 2u_{tt} = 0$
   - vii) $u_{xy} + u_{xz} + u_{xt} + u_{zt} = 0$
   - viii) $u_{xx} + 2u_{xy} - 2u_{xz} - 4u_{yz} + 2u_{yt} + u_{zz} = 0$
   - ix) $u_{xx} + 2u_{xz} - 2u_{xt} + u_{yy} + 2u_{yz} + 2u_{yt} + 2u_{zz} + 2u_{tt} = 0$
2. Find a normal form for the following partial differential equations:
   - i) $u_{x_1x_1} + 2\sum_{k=2}^{n} u_{x_kx_k} - 2\sum_{k=1}^{n-1} u_{x_kx_{k+1}} = 0$
   - ii) $u_{x_1x_1} - 2\sum_{k=2}^{n}(-1)^k u_{x_{k-1}x_k} = 0$
   - iii) $\sum_{k=1}^{n} k u_{x_kx_k} + 2\sum_{l<k} u_{x_lx_k} = 0$
   - iv) $\sum_{k=1}^{n} u_{x_kx_k} + \sum_{l<k} u_{x_lx_k} = 0$
   - v) $\sum_{l<k} u_{x_lx_k} = 0$

9.5 Reduced normal form

Suppose now that the operator L in the previous section has *constant coefficients* and it is *linear*, that is, it has the form

$$Lu(x) = \mathrm{Tr}\,\big(A \cdot u''(x)\big) + \langle b, u'(x)\rangle + c\,u(x)\,,$$

where A is a real symmetric matrix of type $n \times n$, b is an element in $\mathbb{R}^n$, and c is a real number. Here, as before, $\langle , \rangle$ denotes the inner product in $\mathbb{R}^n$. Since A is constant, this operator has the same type at every point of $\mathbb{R}^n$. We also suppose that we have transformed L to normal form as we did it in the previous section, that is, A is a diagonal matrix hence L has the form

$$Lu(x) = \sum_{k=1}^{n} \lambda_k\, \partial_k^2 u(x) + \sum_{k=1}^{n} b_k\, \partial_k u(x) + c\,u(x)\,. \tag{9.2}$$

Here $\lambda_1, \lambda_2, \ldots, \lambda_n$ are all the real eigenvalues of the matrix A, while $b_1, b_2, \ldots, b_n$, and c are real numbers. Now our purpose is to find further transformations which convert L into even simpler, so-called *reduced normal form*. This means that we are looking for an appropriate transformation such that in the new operator K the coefficients of the second order derivatives are 0 and ± 1, further K does not contain either the first derivatives, or the function itself.

The first requirement can be satisfied with the following choice: let for $k = 1, 2, \ldots, n$

$$y_k = \begin{cases} x_k/\sqrt{|\lambda_k|} & \text{if } \lambda_k \neq 0, \\ x_k & \text{otherwise.} \end{cases}$$

Let further $v(y) = u(x)$, where $y = (y_1, y_2, \ldots, y_n)$, then

$$\partial_k\, u(x) = \begin{cases} \big(1/\sqrt{|\lambda_k|}\big) \cdot \partial_k v(y) & \text{if } \lambda_k \neq 0, \\ \partial_k v(y) & \text{otherwise}\,, \end{cases}$$

further

$$\partial_k^2\, u(x) = \begin{cases} \big(1/|\lambda_k|\big) \cdot \partial_k v(y) & \text{if } \lambda_k \neq 0, \\ \partial_k^2 v(y) & \text{otherwise}\,. \end{cases}$$

Then, by substitution, we have

$$Lu(x) = Kv(y) = \sum_{k=1}^{n} \operatorname{sgn} \lambda_k\, \partial_k^2 v(y) + \sum_{k=1}^{n} c_k\, \partial_k v(y) + c\,v(y)\,,$$

where

$$c_k = b_k \frac{1}{\sqrt{|\lambda_k|}}$$

whenever $k = 1, 2, \ldots, n$. It follows that in K the coefficients of the second order derivatives are 0, 1 or -1.

The next step is to define

$$w(y) = v(y)\, e^{-\sum_{i=1}^n \alpha_i y_i},$$

where we will choose the real numbers $\alpha_1, \alpha_2, \ldots, \alpha_n$ in a reasonable way such that K has an even simpler form. In any case we obtain

$$v(y) = w(y)\, e^{\sum_{i=1}^n \alpha_i y_i},$$

$$\partial_k\, v(y) = \big(\partial_k\, w(y) + \alpha_k\, w(y)\big)\, e^{\sum_{i=1}^n \alpha_i y_i},$$

$$\partial_k^2\, v(y) = \big(\partial_k^2\, w(y) + 2\alpha_k\, \partial_k\, w(y) + \alpha_k^2\, w(y)\big)\, e^{\sum_{i=1}^n \alpha_i y_i},$$

whence

$$Mw(y) = e^{-\sum_{i=1}^n \alpha_i y_i} Kv(y) = \sum_{k=1}^n \operatorname{sgn} \lambda_k \partial_k^2 w(y)$$

$$+ \sum_{k=1}^n (c_k + 2\alpha_k \operatorname{sgn} \lambda_k)\partial_k w(y) + \Big[c + \sum_{k=1}^n (\alpha_k^2 \operatorname{sgn} \lambda_k + c_k \alpha_k)\Big]\, w(y)\,.$$

Now for those values of k for which $\lambda_k \neq 0$ we let

$$\alpha_k = -\frac{c_k}{2 \operatorname{sgn} \lambda_k}\,.$$

Then M has the form

$$Mw(y) = \sum_{\lambda_k \neq 0} \operatorname{sgn} \lambda_k\, \partial_k^2\, w(y) + \sum_{\lambda_k = 0} c_k\, \partial_k w(y) + d\, w(y)\,,$$

where

$$d = c - \frac{1}{4} \sum_{\lambda_k \neq 0} c_k^2 \operatorname{sgn} \lambda_k + \sum_{\lambda_k = 0} c_k \alpha_k\,.$$

In order to choose the other parameters α we assume first that there is a k_0 such that $\lambda_{k_0} = 0$, and $c_{k_0} \neq 0$. Then we define

$$\alpha_k = \begin{cases} 0 & \text{if } \lambda_k = 0 \text{ and } k \neq k_0, \\ (1/4c_{k_0}) \cdot \sum_{\lambda_k \neq 0} c_k^2 \operatorname{sgn} \lambda_k - c/c_{k_0} & \text{if } k = k_0\,. \end{cases}$$

Then we have

$$Mw(y) = \sum_{\lambda_k \neq 0} \operatorname{sgn} \lambda_k \, \partial_k^2 \, w(y) + \sum_{\lambda_k = 0} c_k \, \partial_k w(y) \, .$$

This means that in M the unknown function does not occur. We can make a further simplification applying the following linear transformation: let for $k = 1, 2, \ldots, n$

$$z_k = \begin{cases} y_k / c_k & \text{if } c_k \neq 0 \text{ and } \lambda_k = 0, \\ y_k & \text{if } \lambda_k \neq 0, \text{ or } c_k = 0 \text{ and } \lambda_k = 0 \, , \end{cases}$$

moreover, with the notation $z = (z_1, z_2, \ldots, z_n)$ we define $\widetilde{w}(z) = w(y)$. As the terms corresponding to the values of k with $\lambda_k = 0$ do not occur in the principal part of the operator, hence the principal part will not change, and we obtain

$$N\tilde{w}(z) = Mw(y) = \sum_{\lambda_k \neq 0} \operatorname{sgn} \lambda_k \, \partial_k^2 \, \tilde{w}(z) + \sum_{\lambda_k = 0} \partial_k \, \tilde{w}(z) \, .$$

This is one of the possible reduced normal forms.

In the second case for each value of k with $\lambda_k = 0$ we also have $c_k = 0$, and for all these values of k we let $\alpha_k = 0$. Then the operator will have the form

$$Mw(y) = \sum_{\lambda_k \neq 0} \operatorname{sgn} \lambda_k \, \partial_{k}^2, w(y) + d \, w(y) \, ,$$

which is the second possible reduced normal form.

We can summarize our results. Let $D \subseteq \mathbb{R}^n$ be a non-empty open set, and let

$$Lu = f(x)$$

be a linear partial differential equation with constant coefficients, where $f : D \to \mathbb{R}$ is a continuous function. Then, by applying the above transformations, we arrive at either of the following two reduced normal forms:

$$\sum_{k=1}^{n} \epsilon_k \, \partial_k^2 \, u + \sum_{\epsilon_k = 0} \partial_k \, u = f(x) \, ,$$

or

$$\sum_{k=1}^{n} \epsilon_k \, \partial_k^2 \, u + c \, u = f(x) \, ,$$

where c is a real number, and the numbers ϵ_k take the valuse $0, 1, -1$ independently. For instance, in the elliptic case all eigenvalues of the original equation has the same sign, and if we suppose that they are positive, then the reduced normal form is

$$\sum_{k=1}^{n} \partial_k^2 u + c u = f(x) .$$

In the hyperbolic case suppose, for instance, that the first eigenvalue is positive and all the others are negative. Then we have the following reduced normal form:

$$\partial_1^2 u - \sum_{k=2}^{n} \partial_k^2 u + c u = f(x) .$$

Finally, in the parabolic case, if the first eigenvalue is zero and all the others are negative, then we have one of the following two forms:

$$\sum_{k=2}^{n} \partial_k^2 u - \partial_1 u = f(x) ,$$

or

$$\sum_{k=2}^{n} \partial_k^2 u + c u = f(x) .$$

These are the most important reduced normal forms in the applications.

9.6 Problems

Find a reduced normal form for the partial differential equations:

1.

$$u_{xy} + a u_x + b u_y + c u = 0 ,$$

where a, b, c are arbitrary constants.

2.

$$u_{xy} - u_x + u_y + 2u = 0 ,$$

3.

$$u_{xx} + 2u_{xy} - 2u_{xz} + 2u_{yy} + 2u_{zz} = 0 .$$

9.7 Normal form in two variables

Let $\Omega \subseteq \mathbb{R}^2 \times \mathbb{R} \times \mathbb{R}^2$, $D \subseteq \mathbb{R}^2$ be non-empty open sets, and let $f : \Omega \to \mathbb{R}$, $a, b, c : D \to \mathbb{R}$ continuous functions such that the determinant

$$d(x,y) = \det \begin{pmatrix} a(x,y) & b(x,y) \\ b(x,y) & c(x,y) \end{pmatrix}$$

has a constant sign on D. For each function u in $\mathcal{C}^2(D)$ and for all (x, y) in D we define

$$Lu(x,y) = a(x,y)\partial_1^2 u(x,y) + 2b(x,y)\partial_1\partial_2 u(x,y) +$$
$$c(x,y)\partial_2^2 u(x,y) + f\big(x, y, u(x,y), u'(x,y)\big)$$

which is a two-dimensional second order differential operator with linear principal part. Obviously, the type of L is the same at each point of D: it is elliptic if $\operatorname{sgn} d = 1$, parabolic if $\operatorname{sgn} d = 0$, and hyperbolic for $\operatorname{sgn} d = -1$. We note that if we do not assume that $\operatorname{sgn} d$ is constant on D, then, by the continuity of the functions a, b, c, with the exception of the parabolic case the operator L has the same type in a neighborhood of each point of D. Our purpose is to find a twice differentiable one-to-one function $\Phi : D \to \mathbb{R}^2$ such that $\det \Phi' \neq 0$, and the mapping $(\xi, \eta) = \Phi(x, y)$ transforms the operator L into normal form on the whole set $\Phi(D)$. It is easy to see that after the transformation the new operator K will have the same type as L. Indeed, if p, q, r denote the coefficient functions of the operator K corresponding to a, b, c, then the determinant of K is

$$d\big(\Phi^{-1}(\xi,\eta)\big) \cdot \Big[\det \Phi'\big(\Phi^{-1}(\xi,\eta)\big)\Big]^2,$$

and if for each (ξ, η) in $\Phi(D)$ we have $v(\xi,\eta) = u\big(\Phi^{-1}(\xi,\eta)\big)$, then

$$Kv(\xi,\eta) = Lu(x,y),$$

further

$$\begin{aligned} p(\xi,\eta) &= \big[a(\partial_1\Phi_1)^2 + 2b\,\partial_1\Phi_1\,\partial_2\Phi_1 + c(\partial_2\Phi_1)^2\big](x,y), \\ q(\xi,\eta) &= \big[a\,\partial_1\Phi_1\,\partial_1\Phi_2 + b\,(\partial_1\Phi_1\,\partial_2\Phi_2 + \partial_2\Phi_1\,\partial_1\Phi_2) + c\,\partial_2\Phi_1\,\partial_2\Phi_2\big](x,y), \\ r(\xi,\eta) &= \big[a(\partial_1\Phi_2)^2 + 2b\,\partial_1\Phi_2\,\partial_2\Phi_2 + c(\partial_2\Phi_2)^2\big](x,y). \end{aligned}$$

Here Φ_1 and Φ_2 denote the coordinate functions of Φ, and, obviously, $(x, y) = \Phi^{-1}(\xi, \eta)$. The first order partial differential equation

$$a(x,y)\big(\partial_1 v\big)^2 + 2b(x,y)\,\partial_1 v\,\partial_2 v + c(x,y)\big(\partial_2 v\big)^2 = 0 \tag{9.3}$$

is called the *characteristic differential equation* of the partial differential equation $Lu = 0$.

9.8 Hyperbolic equations

Using the notation of the previous section we assume that the operator L is hyperbolic on D, that is, $d(x,y) < 0$ holds for each (x,y) in D. Suppose, moreover, that $a(x,y) \neq 0$ for every (x,y) in D. Then the characteristic equation (9.3) can be written in the following form:

$$\big(\partial_1 v + f_1(x,y)\partial_2 v\big)\big(\partial_1 v + f_2(x,y)\partial_2 v\big) = 0\,,$$

where

$$f_1(x,y) = \frac{b(x,y) - \sqrt{-d(x,y)}}{a(x,y)}\,, \qquad f_2(x,y) = \frac{b(x,y) + \sqrt{-d(x,y)}}{a(x,y)}\,,$$

whenever (x,y) is in D. Assume that φ, respectively ψ are twice continuously differentiable solutions of the partial differential equations

$$\partial_1 v + f_1(x,y)\partial_2 v = 0\,,$$

respectively

$$\partial_1 v + f_2(x,y)\partial_2 v = 0$$

such that

$$\partial_2\varphi(x,y) \neq 0\,, \qquad \text{respectively} \qquad \partial_2\psi(x,y) \neq 0$$

holds whenever (x,y) is in D. If a, b, c are twice continuously differentiable, then, by the properties of the characteristic function, such solutions exist in some neighborhood of each point of D. We let

$$\Phi = (\varphi, \psi)\,,$$

then

$$\det \Phi'(x,y) = \big(f_1(x,y) - f_2(x,y)\big) \cdot \partial_2\varphi(x,y) \cdot \partial_2\psi(x,y) \neq 0$$

holds for each point (x,y) in D, hence Φ is locally injective. Suppose that Φ is injective on D, and apply this transformation on the operator L. The coefficient functions of the new operator corresponding to a, b, c will be denoted by p, q, r, respectively. Then the equation $Lu = 0$ turns into

$$2q(\xi,\eta)\,\partial_1\partial_2 v(\xi,\eta) + g\big(\xi,\eta,v(\xi,\eta),v'(\xi,\eta)\big) = 0\,.$$

As the determinant here is $-q(\xi,\eta)^2$, hence $q(\xi,\eta) \neq 0$ for each (ξ,η) in $\Phi(D)$, and after division by $2q(\xi,\eta)$ we get the equation

$$\partial_1\partial_2 v(\xi,\eta) + \frac{1}{2q(\xi,\eta)} g\big(\xi,\eta,v(\xi,\eta),v'(\xi,\eta)\big) = 0\,,$$

which holds for each (ξ, η) in $\Phi(D)$. Here we let $t = \xi + \eta$ and $s = \xi - \eta$, further

$$w(t, s) = v\left(\frac{t+s}{2}, \frac{t-s}{2}\right).$$

Applying this transformation the equation turns into the normal form

$$\partial_1^2 w - \partial_2^2 w + h\big(t, s, w(t, s), w'(t, s)\big) = 0\,,$$

where the variable (t, s) is taken from some open set in $\mathbb{R}^2$, and h is a continuous function which can easily be calculated from f.

We note that if in the original equation $c(x, y) \neq 0$ holds for each (x, y) in D, then we can transform the equation into normal form in a similar way. If, however, for each (x, y) in D we have $a(x, y) = c(x, y) = 0$, then the equation has a form for which we need to apply the second transformation only.

We apply the above method for the partial differential equation

$$x^2 \frac{\partial^2 u}{\partial x^2} - y^2 \frac{\partial^2 u}{\partial y^2} = 0\,.$$

As

$$d(x, y) = \det \begin{pmatrix} x^2 & 0 \\ 0 & -y^2 \end{pmatrix} = -x^2 y^2 < 0$$

for $x \neq 0$, $y \neq 0$, hence the equation is hyperbolic in each quadrant. We consider the case $x, y > 0$. Now we have

$$f_1(x, y) = -\frac{y}{x}, \qquad f_2(x, y) = \frac{y}{x}\,,$$

hence

$$\varphi(x, y) = xy, \qquad \psi(x, y) = \frac{y}{x}\,.$$

By our above considerations we have to apply the following transformation:

$$\xi = xy, \qquad \eta = \frac{y}{x}, \qquad v(\xi, \eta) = u(x, y)\,.$$

We use some simplified notation in the following computation:

$$u_{xx} = v_{\xi\xi} \cdot y^2 + v_{\xi\eta}\left(-2\frac{y^2}{x^2}\right) + v_{\eta\eta} \cdot \frac{y^2}{x^4} + v_\eta \cdot \frac{2y}{x^3}\,,$$

$$u_{yy} = v_{\xi\xi} \cdot x^2 + 2v_{\xi\eta} + v_{\eta\eta} \cdot \frac{1}{x^2}\,,$$

and substitution gives

$$-4\xi\eta\, v_{\xi\eta} + 2\eta\, v_\eta = 0\,,$$

or, after simplification

$$v_{\xi\eta} - \frac{1}{2\xi}\, v_\eta = 0\,. \tag{9.4}$$

Now we apply the transformation $s = \xi + \eta$, $t = \xi - \eta$ we get that the function $w(s,t) = v(\xi,\eta)$ satisfies

$$w_{ss} - w_{tt} - \frac{1}{s+t}(w_s - w_t) = 0\,,$$

which is the normal form

$$\partial_1^2 w - \partial_2^2 w - \frac{1}{s+t}(\partial_1 w - \partial_2 w) = 0\,.$$

We note that if we want to compute the explicit solution, then the form (9.4) is more useful. Indeed, for a fixed η we can solve (9.4) by introducing the function $z(\xi) = \partial_2 v(\xi,\eta)$ which satisfies the following ordinary differential equation:

$$z' - \frac{1}{\xi}\, z = 0\,.$$

Its solution has the form $z(\xi) = c(\eta)\sqrt{\xi}$, where c is an arbitrary function which is, by the properties of z, obviously continuously differentiable. This means that we have for v

$$\partial_2 v(\xi,\eta) = c(\eta)\sqrt{\xi}\,,$$

and, by holding ξ fixed and integrating, it follows

$$v(\xi,\eta) = A(\eta)\sqrt{\xi} + B(\xi)$$

where A, B are arbitrary twice continuously differentiable functions. Finally, using the relation between u and v, we infer

$$u(x,y) = v(\xi,\eta) = A\left(\frac{y}{x}\right)\sqrt{xy} + B(xy)\,,$$

which is the general solution of the original equation, with arbitrary twice continuously differentiable functions A, B.

9.9 Problems

Find a normal form for the following partial differential equations on domains, where the type does not change.

1. $u_{xx} - 2u_{xy} - 3u_{yy} + u_y = 0$
2. $y^2 u_{xx} - x^2 u_{yy} = 0$
3. $4y^2 u_{xx} - e^{2x} u_{yy} = 0$

9.10 Parabolic equations

Now we suppose that the operator L is parabolic on D, that is

$$d(x,y) = 0$$

holds for each (x, y) in D. Suppose, moreover, that $a(x, y) \neq 0$ holds at every point of D, too. Then the characteristic equation (9.3) of the partial differential equation

$$Lu = 0$$

has the form

$$\left(\partial_1 v + \frac{b(x,y)}{a(x,y)}\,\partial_2 v\right)^2 = 0\,,$$

that is

$$\partial_1 v + \frac{b(x,y)}{a(x,y)}\,\partial_2 v = 0\,.$$

Let for each (x, y) in D

$$f(x,y) = \frac{b(x,y)}{a(x,y)}\,,$$

further let φ denote a twice continuously differentiable solution of the partial differential equation

$$\partial_1 v + f(x,y)\partial_2 v = 0$$

which satisfies $\partial_2\varphi(x, y) \neq 0$, whenever (x, y) is in D. Let, moreover, $\psi : D \to \mathbb{R}$ be a twice continuously differentiable function such that the function $\Phi = (\varphi, \psi)$ is injective on D, and $\det \Phi'(x, y) \neq 0$ whenever (x, y) is in D. We apply this transformation on the operator L, then from the equation $Lu = 0$ we get

$$r(\xi,\eta)\,\partial_2^2 v(\xi,\eta) + g\big(\xi,\eta,v(\xi,\eta),v'(\xi,\eta)\big) = 0\,,$$

as the new determinant is $p(\xi, \eta) \cdot r(\xi, \eta) - q(\xi, \eta)^2 = 0$, however, by the choice of Φ we have $p = 0$, hence $q = 0$. On the other hand, if $r(\xi_0, \eta_0) = 0$ at some point $(\xi_0, \eta_0) = \Phi(x_0, y_0)$ of the set $\Phi(D)$, then we have

$$\partial_1\varphi(x_0,y_0) + f(x_0,y_0)\,\partial_2\varphi(x_0,y_0) = 0\,,$$
$$\partial_1\psi(x_0,y_0) + f(x_0,y_0)\,\partial_2\psi(x_0,y_0) = 0\,,$$

which is impossible, as the determinant of this system of equations is nonzero. We can divide by $r(\xi, \eta)$ to obtain the normal form

$$\partial_2^2 v + \frac{1}{r(\xi,\eta)} g\big(\xi,\eta,v(\xi,\eta),v'(\xi,\eta)\big) = 0\,.$$

If we have $c(x,y) \neq 0$ at each point (x,y) in D, then we can proceed similarly. Finally, by parabolicity, the equality $a(x,y) = c(x,y) = 0$ cannot hold at each (x,y) in D.

As an illustration we consider the partial differential equation

$$x^2 \frac{\partial^2 u}{\partial x^2} + 2xy \frac{\partial^2 u}{\partial x \partial y} + y^2 \frac{\partial^2 u}{\partial y^2} = 0 .$$

Since we have

$$\det \begin{pmatrix} x^2 & xy \\ xy & y^2 \end{pmatrix} = 0 ,$$

hence the equation is parabolic on the whole plane. We consider the region $x > 0$. The characteristic equation has the form

$$\partial_1 v + \frac{y}{x} \partial_2 v = 0 ,$$

and a solution is a non-zero first integral of the differential equation

$$y' = \frac{y}{x} .$$

As $(y/x)' = 0$, hence we get $y/x =$ constant, and we conclude that a first integral is

$$\varphi(x,y) = \frac{y}{x} .$$

Now we look for a twice continuously differentiable function ψ such that the determinant

$$\det \begin{pmatrix} \varphi_x & \varphi_y \\ \psi_x & \psi_y \end{pmatrix}$$

never vanishes. Let, for instance, $\psi(x,y) = x$, and we apply the transformation $\xi = x$, $\eta = \frac{y}{x}$. This brings the equation into the normal form

$$\frac{\partial^2 v}{\partial \xi^2} = 0 .$$

9.11 Problems

Find a normal form for the following partial differential equations on domains, where its type does not change:

1. $4u_{xx} - 4u_{xy} + u_{yy} - 2u_y = 0$
2. $x^2 u_{xx} - 2xu_{xy} + u_{yy} = 0$

9.12 Elliptic equations

Finally, we suppose that the operator L is elliptic on D, that is

$$d(x,y) > 0$$

holds for each (x,y) in D. Then $a(x,y) \neq 0$ for each (x,y) in D. The characteristic equation of the partial differential equation $Lu = 0$ can be written in the form

$$\left(\partial_1 v + \frac{b(x,y) - i\sqrt{d(x,y)}}{a(x,y)}\,\partial_2 v\right)\left(\partial_1 v + \frac{b(x,y) + i\sqrt{d(x,y)}}{a(x,y)}\,\partial_2 v\right) = 0\,.$$

Let

$$f_0(x,y) = \frac{b(x,y) + i\sqrt{d(x,y)}}{a(x,y)}$$

whenever (x,y) is in D, and let ζ be a twice continuously differentiable complex solution of the partial differential equation

$$\partial_1 w + f_0(x,y)\partial_2 w = 0$$

such that $\partial_2\zeta(x,y) \neq 0$ for each (x,y) in D. If a, b, c are analytic functions, then, by Theorem 8.2.1 of Kovalevskaya, every point of D has a neighborhood in which such a solution exists. We let

$$\varphi = \operatorname{Re}\zeta, \qquad \psi = \operatorname{Im}\zeta, \qquad \Phi = (\varphi, \psi)\,.$$

It follows

$$\det\Phi'(x,y) = -\frac{\sqrt{d(x,y)}}{a(x,y)}\left[\big(\partial_2\varphi(x,y)\big)^2 + \big(\partial_2\psi(x,y)\big)^2\right] \neq 0\,,$$

hence Φ is locally injective. For the sake of simplicity we assume that Φ is injective on D, and we apply this transformation on L. Then we obtain the following:

$$p(\xi,\eta)\,\partial_1^2 v(\xi,\eta) + p(\xi,\eta)\,\partial_2^2 v(\xi,\eta) + g\big(\xi,\eta,v(\xi,\eta),v'(\xi,\eta)\big) = 0\,.$$

Here the determinant is $p(\xi,\eta)^2 > 0$, hence $p(\xi,\eta) \neq 0$ at every point (ξ,η) in $\Phi(D)$. After division we arrive at the normal form

$$\partial_1^2 v + \partial_2^2 v + \frac{1}{p(\xi,\eta)} g\big(\xi,\eta,v(\xi,\eta),v'(\xi,\eta)\big) = 0\,.$$

We apply this method for the partial differential equation

$$(1+x^2)\,\frac{\partial^2 u}{\partial x^2} + (1+y^2)\,\frac{\partial^2 u}{\partial y^2} + x\,\frac{\partial u}{\partial x} + y\,\frac{\partial u}{\partial y} = 0\,.$$

By the relation

$$d(x,y) = \det \begin{pmatrix} 1+x^2 & 0 \\ 0 & 1+y^2 \end{pmatrix} = (1+x^2)(1+y^2) > 0$$

the equation is elliptic on the whole plane. The characteristic equation is

$$\partial_1 w + i\frac{\sqrt{(1+x^2)(1+y^2)}}{1+x^2}\,\partial_2 w = 0\,,$$

that is

$$\partial_1 w + i\frac{\sqrt{1+y^2}}{\sqrt{1+x^2}}\,\partial_2 w = 0\,.$$

One solution of this equation is given by a first integral of the ordinary differential equation

$$y' = i\frac{\sqrt{1+y^2}}{\sqrt{1+x^2}}\,.$$

This equation is separable, and a first integral is given by

$$\zeta(x,y) = \operatorname{arsh} y - i \operatorname{arsh} x\,,$$

whose real and imaginary parts will serve as the components of the necessary transformation. Let

$$\xi = \operatorname{arsh} x, \qquad \eta = \operatorname{arsh} y\,,$$

then we have

$$x = \operatorname{sh}\xi, \qquad y = \operatorname{sh}\eta\,,$$

and applying the transformation we obtain the normal form

$$\frac{\partial^2 v}{\partial \xi^2} + \frac{\partial^2 v}{\partial \eta^2} = 0\,.$$

This is the Laplace equation.

9.13 Problems

1. Find a normal form for the following partial differential equations on domains, where the type does not change.

   i) $u_{xx} - 6u_{xy} + 10u_{yy} + u_x - 3u_y = 0$
   ii) $x^2 u_{xx} + y^2 u_{yy} = 0$
   iii) $y^2 u_{xx} + x^2 u_{yy} = 0$
   iv) $(1+x^2)u_{xx} + (1+y^2)u_{yy} + yu_y = 0$
   v) $u_{xx} - 2\sin x u_{xy} + (2-\cos^2 x)u_{yy} = 0$

2. Solve the following problems.

   i) Find the general solutions of the following partial differential equation:

$$x^2 \frac{\partial^2 u}{\partial x^2} - y^2 \frac{\partial^2 u}{\partial y^2} = 0 \,.$$

   ii) Find a normal form for the following partial differential equation:

$$(1 + x^2)\, \frac{\partial^2 u}{\partial x^2} + (1 + y^2)\, \frac{\partial^2 u}{\partial y^2} + x\, \frac{\partial u}{\partial x} + y\, \frac{\partial u}{\partial y} = 0 \,.$$

   iii) Find the general solutions of the following partial differential equation:

$$x^2 \frac{\partial^2 u}{\partial x^2} + 2xy \frac{\partial^2 u}{\partial x \partial y} + y^2 \frac{\partial^2 u}{\partial y^2} = 0 \,.$$

3. Find a normal form for the following partial differential equations:

   i)

$$e^{2x}\, \frac{\partial^2 u}{\partial x^2} + 2e^{x+y}\, \frac{\partial^2 u}{\partial x \partial y} + e^{2y}\, \frac{\partial^2 u}{\partial y^2} = 0,$$

   ii)

$$\frac{\partial^2 u}{\partial x^2} + (1 + y)^2 \frac{\partial^2 u}{\partial y^2} = 0,$$

   iii)

$$y^2\, \frac{\partial^2 u}{\partial x^2} + x^2\, \frac{\partial^2 u}{\partial y^2} = 0,$$

   iv)

$$y^2\, \frac{\partial^2 u}{\partial x^2} - e^{2x}\, \frac{\partial^2 u}{\partial y^2} + \frac{\partial u}{\partial x} = 0,$$

   v)

$$\frac{\partial^2 u}{\partial x^2} + 5\, \frac{\partial^2 u}{\partial x \partial y} + 6\, \frac{\partial^2 u}{\partial y^2} + 3\, \frac{\partial u}{\partial x} + 2\, \frac{\partial u}{\partial y} = 0,$$

   vi)

$$\frac{\partial^2 u}{\partial x^2} + y\, \frac{\partial^2 u}{\partial y^2} = 0,$$

   vii)

$$\frac{\partial^2 u}{\partial x^2} + xy\, \frac{\partial^2 u}{\partial y^2} = 0.$$

4. Find the general solutions of the following partial differential equations:

   i)
$$\frac{\partial^2 u}{\partial x^2} - \frac{\partial^2 u}{\partial y^2} = 0,$$

   ii)
$$\frac{\partial^2 u}{\partial x^2} + 2\,\frac{\partial^2 u}{\partial x \partial y} - 3\,\frac{\partial^2 u}{\partial y^2} + 2\,\frac{\partial u}{\partial x} + 6\,\frac{\partial u}{\partial y} = 0,$$

   iii)
$$\frac{\partial^2 u}{\partial x^2} - \frac{\partial^2 u}{\partial y^2} + \frac{1}{2}\left(\frac{\partial u}{\partial x} + \frac{\partial u}{\partial y}\right) = 0,$$

   iv)
$$(x-y)\,\frac{\partial^2 u}{\partial x^2} + (xy - y^2 - x + y)\,\frac{\partial^2 u}{\partial x \partial y} = 0.$$

5. Find a normal form for the following partial differential equations on domains, where the type does not change.

   i) $u_{xx} - xu_{yy} = 0$
   ii) $u_{xx} - yu_{yy} = 0$
   iii) $xu_{xx} - yu_{yy} = 0$
   iv) $yu_{xx} - xu_{yy} = 0$

6. Find the general solution of the following partial differential equations.

   i) $u_{xy} = 0$
   ii) $u_{xx} - a^2 u_{yy} = 0$
   iii) $u_{xx} - 2u_{xy} - 3u_{yy} = 0$
   iv) $u_{xy} + au_x = 0$
   v) $3u_{xx} - 5u_{xy} - 2u_{yy} + 3u_x + u_y = 2$
   vi) $u_{xy} + au_x + bu_y + abu = 0$
   vii) $u_{xy} - 2u_x - 3u_y + 6u = 2e^{x+y}$
   viii) $u_{xx} + 2au_{xy} + a^2 u_{yy} + u_x + au_y = 0$

7. Find the general solution of the following partial differential equations on domains, where the type does not change.

   i) $yu_{xx} + (x-y)u_{xy} - xu_{yy} = 0$
   ii) $x^2 u_{xx} - y^2 u_{yy} = 0$
   iii) $x^2 u_{xx} + 2xyu_{xy} - 3y^2 u_{yy} - 2xu_x = 0$
   iv) $x^2 u_{xx} + 2xyu_{xy} + y^2 u_{yy} = 0$
   v) $u_{xy} - xu_x + u = 0$

vi) $u_{xy} + 2xyu_y - 2xu = 0$
vii) $u_{xy} + u_x + yu_y + (y-1)u = 0$
viii) $u_{xy} + xu_x + 2yu_y + 2xyu = 0$

Chapter 10

SPECIAL PROBLEMS IN TWO VARIABLES

10.1 Goursat problem for hyperbolic equations

Let $I, J \subseteq \mathbb{R}$ be non-empty bounded open intervals with left endpoint 0. Let $D = I \times J$ and let $a, b, c : D^{cl} \to \mathbb{R}$, $\varphi : I^{cl} \to \mathbb{R}$ and $\psi : J^{cl} \to \mathbb{R}$ be continuous functions such that $\varphi(0) = \psi(0)$ holds. We consider on D the hyperbolic linear partial differential equation

$$\partial_1 \partial_2 u + a(x,y)\,\partial_1 u + b(x,y)\partial_2 u + c(x,y)u = f(x,y)\,, \tag{10.1}$$

which is written in normal form. We consider the following problem which is called the *Goursat problem* for equation (10.1): find a continuously differentiable function $u : D^{cl} \to \mathbb{R}$ which is a solution of (10.1) on D, further we have

$$u(x,0) = \varphi(x), \qquad u(0,y) = \psi(y) \tag{10.2}$$

for each x in I^{cl} and y in J^{cl}.

To solve the Goursat problem (10.1), (10.2) let $\mathcal{C}^1(D^{cl})$ denote the Banach space of all continuously differentiable functions u on D^{cl} equipped with the norm

$$\|u\| = \max|u(x,y)| + \max|\partial_1 u(x,y)| + \max|\partial_2 u(x,y)|\,,$$

where the maximum is taken over the elements (x,y) in D^{cl}. Moreover, for each u in $\mathcal{C}^1(D^{cl})$ we define

$$Au(x,y) \tag{10.3}$$

$$= \int_0^x \int_0^y \big[a(\xi,\eta)\partial_1 u(\xi,\eta) + b(\xi,\eta)\partial_2 u(\xi,\eta) + c(\xi,\eta)\,u(\xi,\eta)\big]\,d\xi\,d\eta\,,$$

where (x,y) is arbitrary in D^{cl}. Then $A : \mathcal{C}^1(D^{cl}) \to \mathcal{C}^1(D^{cl})$ is a bounded linear operator.

Theorem 10.1.1. *Let $\varphi : I^{cl} \to \mathbb{R}$, $\psi : J^{cl} \to \mathbb{R}$ be continuously differentiable functions, further for each (x, y) in D^{cl} we let*

$$g(x,y) = \varphi(x) + \psi(y) - \varphi(0) + \int_0^x \int_0^y f(\xi,\eta)\, d\xi\, d\eta\,. \tag{10.4}$$

The function u in $\mathcal{C}^1(D^{cl})$ is a solution of the Goursat problem (10.1), (10.2) *if and only if*

$$u + Au = g \tag{10.5}$$

holds on D^{cl}.

Proof. Suppose that the function u in $\mathcal{C}^1(D^{cl})$ is a solution of the Goursat problem (10.1), (10.2). Then for each (ξ, η) in D^{cl} it follows

$$\partial_1 \partial_2 u(\xi,\eta) + a(\xi,\eta)\, \partial_1 u(\xi,\eta) + b(\xi,\eta)\partial_2 u(\xi,\eta) + c(\xi,\eta)u(\xi,\eta) = f(\xi,\eta)\,.$$

We integrate over the rectangle $[0, x] \times [0, y]$, where (x, y) is in D^{cl}:

$$\int_0^x \int_0^y \partial_1 \partial_2 u(\xi,\eta)\, d\xi\, d\eta + Au(x,y) = \int_0^x \int_0^y f(\xi,\eta)\, d\xi\, d\eta\,.$$

On the other hand

$$\int_0^x \int_0^y \partial_1 \partial_2 u(\xi,\eta)\, d\xi\, d\eta \int_0^x \left[\int_0^y \partial_1 \partial_2 u(\xi,\eta)\, d\eta\right] d\xi$$

$$= \int_0^x \left[\partial_1 u(\xi,\eta)\right]_0^y d\xi = \int_0^x \left[\partial_1 u(\xi,y) - \partial_1 u(\xi,0)\right] d\xi = \left[u(\xi,y) - u(\xi,0)\right]_0^x$$

$$= u(x,y) - u(x,0) - u(0,y) + u(0,0) = u(x,y) - \varphi(x) - \psi(y) + \varphi(0)\,,$$

which implies the necessity of the condition.

To prove the converse we assume that u is in $\mathcal{C}^1(D^{cl})$ such that (10.5) holds on D^{cl}. Then, by $u = g - Au$ we have that the partial derivative $\partial_1\partial_2 u$ exists and

$$\partial_1\partial_2 u = \partial_1\partial_2 g - \partial_1\partial_2(Au) = f - (a\partial_1 u + b\partial_2 u + cu)$$

on D, hence u is a solution of (10.1). On the other hand, (10.2) is obvious. □

Theorem 10.1.2. *Let g be arbitrary in $\mathcal{C}^1(D^{cl})$. Then the series*

$$\sum (-1)^n A^n g$$

is convergent in the Banach space $\mathcal{C}^1(D^{cl})$, and its sum u satisfies (10.5).

Proof. It is easy to prove by induction that for each g in $\mathcal{C}^1(D^{cl})$ we have

$$\|A^n g\| \leqslant 3\,\frac{K^n}{n!}\,\|g\|\,,$$

where $K = 2Ml(l+2)$ with

$$M = \max_{(x,y)\in D^{cl}} \{|a(x,y)|, |b(x,y)|, |c(x,y)|\}\,,$$

further I and J are contained in the interval $[0,l]$. This implies that the above series is absolutely convergent in $\mathcal{C}^1(D^{cl})$, hence it is convergent. Finally, if u denotes the sum of the series, then

$$Au = \sum_{n=0}^{\infty} (-1)^n A^{n+1} g\,,$$

which implies $u + Au = g$. □

Theorem 10.1.3. *Let the coefficient functions in* (10.1) *be continuous on* D^{cl}, *let* $\varphi : I^{cl} \to \mathbb{R}$ *and* $\psi : J^{cl} \to \mathbb{R}$ *be continuously differentiable functions, further let* $\varphi(0) = \psi(0)$. *Then the Goursat problem* (10.1), (10.2) *has a unique solution on* D^{cl}.

Proof. The existence of the solution follows from the previous two theorems. To prove the uniqueness we note the estimate obtained for $\|A^n\|$ in the proof of the previous theorem implies that the operator $E + A$ is invertible on the space $\mathcal{C}^1(D^{cl})$ which gives our statement. Here E is the identity operator on $\mathcal{C}^1(D^{cl})$. □

As an application we solve the Goursat problem

$$\begin{aligned} u_{xy} + c\,u &= 0\,,\\ u(x,0) &= 1\,,\\ u(0,y) &= 1 \end{aligned}$$

on the unit square $[0,1] \times [0,1]$, where c is a real number.

Using the above results the solution u of the Goursat problem is also a solution of the integro-differential equation

$$u + Au = g\,,$$

where

$$Au(x,y) = c \int_0^x \int_0^y u(\xi,\eta)\,d\xi\,d\eta\,,$$

and $g(x,y) = 1$. This is a simple integral equation with the unique solution

$$u = \sum_{n=0}^{\infty} (-1)^n A^n g\,,$$

and here $A^0 g = g = 1$. It follows

$$Ag(x,y) = c\int_0^x \int_0^y d\xi\, d\eta = c\,x\,y\,,$$

$$A^2 g(x,y) = c\int_0^x \int_0^y c\xi\eta\, d\xi\, d\eta = \frac{(c\,x\,y)^2}{(2!)^2}\,,$$

or, more generally

$$A^n g(x,y) = \frac{(c\,x\,y)^n}{(n!)^2}\,,$$

hence the unique solution of the problem is

$$u(x,y) = \sum_{n=0}^{\infty} \frac{(c\,x\,y)^n}{(n!)^2}\,.$$

We note that similar results can be proved on more general domain D, and also the solution method works in those cases. For instance, $D \subseteq \mathbb{R}^2$ may have the form

$$D = \{(x,y) \,|\, \alpha < x < \beta, p(x) < y < q(x)\}\,,$$

where $\alpha < \beta$ are extended real numbers, that is, they can take the values $\pm\infty$, and the functions $p, q :]\alpha, \beta[\to \mathbb{R}$ are continuously differentiable. In this case the boundary conditions have the form $u\big(x, p(x)\big) = \varphi(x), u\big(x, q(x)\big) = \psi(x)$ for $a < x < b$, and obviously φ, ψ are defined on the open interval $]\alpha, \beta[$. Also, the role of x and y can be interchanged in this modified setting. Examples can be found in the following problem section.

10.2 Problems

Solve the following Goursat problems.

1.

$$\begin{aligned} u_{xy} + u_x &= x, \qquad (x, y > 0) \\ u(0,y) &= y^2 \\ u(x,0) &= x^2 \end{aligned}$$

2.

$$
\begin{aligned}
u_{xy} + x^2 y u_x &= 0, \qquad (x, y > 0) \\
u(0, y) &= 0 \\
u(x, 0) &= x
\end{aligned}
$$

3.

$$
\begin{aligned}
2u_{xx} - 2u_{yy} + u_x + u_y &= 0, \qquad (y > |x|) \\
u(x, x) &= 1 \\
u(x, -x) &= (x + 1)e^x
\end{aligned}
$$

4.

$$
\begin{aligned}
2u_{xx} + u_{xy} - u_{yy} + u_x + u_y &= 0, \qquad \Big(-\frac{1}{2}x < y < x, x > 0\Big) \\
u(x, x) &= 1 + 3x \\
u\Big(x, -\frac{1}{2}x\Big) &= 1
\end{aligned}
$$

5.

$$
\begin{aligned}
u_{xy} - e^x u_{yy} &= 0, \qquad (-e^x < y < x, x > 0) \\
u(0, y) &= y^2 \\
u(x, -e^x) &= 1 + x^2
\end{aligned}
$$

6.

$$
\begin{aligned}
u_{xx} - u_{yy} &= x^2 - y^2, \qquad (y < |x|) \\
u(x, x) &= x^2 \\
u(x, -x) &= -x^2
\end{aligned}
$$

10.3 Cauchy problem for hyperbolic equations

Let $-\infty \leqslant \alpha < \beta \leqslant +\infty$ be real numbers, and let $\varphi :]\alpha, \beta[\to \mathbb{R}$ be a continuously differentiable function such that $\varphi'(x) > 0$ holds whenever $\alpha < x < \beta$. Let further

$$
\gamma = \lim_{x \to \alpha+} \varphi(x), \qquad \delta = \lim_{x \to \beta-} \varphi(x) ,
$$

and $D =]\alpha, \beta[\times]\gamma, \delta[$. Given the continuous functions $a, b, c, f : D \to \mathbb{R}$ and the continuously differentiable functions $\psi, \chi :]\alpha, \beta[\to \mathbb{R}$ we consider on D the linear hyperbolic partial differential equation of the following normal form

$$
\partial_1 \partial_2 u + a(x, y)\partial_1 u + b(x, y)\partial_2 u + c(x, y)u = f(x, y) . \tag{10.6}
$$

In particular, we shall consider the following problem which is called the *Cauchy problem* for equation (10.6): find a continuously differentiable function $u : D \to \mathbb{R}$ such that u is a solution of (10.6) on D and it satisfies

$$u\big(x, \varphi(x)\big) = \psi(x), \qquad \partial_2 u\big(x, \varphi(x)\big) = \chi(x) \tag{10.7}$$

for each x in $]\alpha, \beta[$. We note that, in general, φ can be replaced by any continuous monotonic function, and the initial conditions can be modified in the way that they prescribe the value of u and some of its directional derivatives along φ.

Theorem 10.3.1. *Under the above conditions the Cauchy problem* (10.6), (10.7) *has a unique twice continuously differentiable solution on* D.

Proof. Let $u : D \to \mathbb{R}$ be a solution of the Cauchy problem (10.6), (10.7) on D, further let

$$\begin{aligned}
v(x,y) &= \partial_1 u(x,y)\,,\\
w(x,y) &= \partial_2 u(x,y)\,,\\
g(x,y) &= f(x,y) - a(x,y)v(x,y) - b(x,y)w(x,y) - c(x,y)u(x,y)\,,
\end{aligned}$$

whenever (x, y) is in D. Then we have

$$\partial_2 u(x,y) = w(x,y), \qquad u\big(x,\varphi(x)\big) = \psi(x)\,,$$

$$\partial_2 v(x,y) = g(x,y), \qquad v\big(x,\varphi(x)\big) = \psi'(x) - \chi(x)\varphi'(x)\,,$$

$$\partial_1 w(x,y) = g(x,y), \qquad w\big(x,\varphi(x)\big) = \chi(x)\,,$$

which implies, by integration:

$$u(x,y) = \psi(x) + \int_{\varphi(x)}^{y} w(x,\eta)\,d\eta\,, \tag{10.8}$$

$$v(x,y) = \psi'(x) - \chi(x)\varphi'(x) + \int_{\varphi(x)}^{y} g(x,\eta)\,d\eta\,, \tag{10.9}$$

$$w(x,y) = \chi\big(\varphi^{-1}(y)\big) + \int_{\varphi^{-1}(y)}^{x} g(\xi,y)\,d\xi\,. \tag{10.10}$$

Conversely, if the continuous functions $u, v, w :]\alpha, \beta[\to \mathbb{R}$ satisfy the integral equation (10.8), then u is a twice continuously differentiable solution of the Cauchy problem (10.6), (10.7). Indeed, we have that

$$\partial_1\partial_2 u(x,y) = f(x,y) - a(x,y)v(x,y) - b(x,y)w(x,y) - c(x,y)u(x,y)\,,$$

$$\partial_2 u(x,y) = w(x,y)\,,$$

and

$$\partial_1 u(x,y) = \psi'(x) + \int_{\varphi(x)}^{y} \partial_1 w(x,\eta)\, d\eta - w\big(x,\varphi(x)\big)\varphi'(x)$$

$$= \psi'(x) - \chi(x)\varphi'(x) + \int_{\varphi(x)}^{y} g(x,\eta)\, d\eta = v(x,y)$$

holds at each point (x, y) in D, since $\partial_1 w = g$.

Consequently, it is enough to show that the integral equation (10.8) has exactly one solution u, v, w on D. First we prove the existence using the method of successive approximation. We introduce the following notation: for each (x, y) in D and for $n = 1, 2, \ldots$ we let

$$\begin{aligned}
u_0(x,y) &= \psi(x)\,,\\
v_0(x,y) &= \psi'(x) - \chi(x)\varphi'(x)\,,\\
w_0(x,y) &= \chi\big(\varphi^{-1}(y)\big)\,,\\
u_n(x,y) &= \psi(x) + \int_{\varphi(x)}^{y} w_{n-1}(x,\eta)\, d\eta\,,\\
v_n(x,y) &= \psi'(x) - \chi(x)\varphi'(x) + \int_{\varphi(x)}^{y} F_{n-1}(x,\eta)\, d\eta\,,\\
w_n(x,y) &= \chi\big(\varphi^{-1}(y)\big) + \int_{\varphi^{-1}(y)}^{x} F_{n-1}(\xi,y)\, d\xi\,,
\end{aligned}$$

where

$$F_k(x,y) = f(x,y) - a(x,y)v_k(x,y) - b(x,y)w_k(x,y) - c(x,y)u_k(x,y)\,,$$

whenever (x, y) is in D, and $k = 0, 1, \ldots$. Applying the same idea used in the proof of the Picard–Lindelöf Theorem 1.7.1 we can prove easily that the function sequences $(u_n)_{n\in\mathbb{N}}$, $(v_n)_{n\in\mathbb{N}}$, $(w_n)_{n\in\mathbb{N}}$ converge uniformly on every compact subinterval of the interval $]\alpha, \beta[$, and if u, v, w denote the respective limit functions, then these are solutions of the integral equation (10.8).

Now we prove the uniqueness. It is easy to see that if $I_1 = [\alpha_1, \beta_1]$ is a compact subinterval of $]\alpha, \beta[$, and

$$K = \max_{x,y\in I}\{|a(x,y)|, |b(x,y)|, |c(x,y)|\}\,,$$

$$d = \varphi(\beta_1) - \varphi(\alpha_1) + \beta_1 - \alpha_1\,,$$

further u_1, v_1, w_1 and u_2, v_2, w_2 are two solutions of the integral equation (10.8), then, with the notation

$$u = u_1 - u_2, \qquad v = v_1 - v_2, \qquad w = w_1 - w_2$$

and

$$B = \max_{x,y \in I_1} \{|u(x,y)|, |v(x,y)|, |w(x,y)|\},$$

we have

$$|u(x,y)|,\ |v(x,y)|,\ |w(x,y)| \leqslant B\,\frac{(dK)^n}{n!}$$

for each x, y in I_1 and for $n = 1, 2, \ldots$. This gives the statement. □

10.4 Problems

Solve the following Cauchy problems.

1.

$$\begin{aligned} u_{xy} - u_{yy} + u_x - u_y &= 0 \\ u(x,0) &= \varphi(x) \\ u_y(x,0) &= \psi(x), \end{aligned}$$

where φ, ψ are continuously differentiable on $\mathbb{R}$.

2.

$$\begin{aligned} x^2 u_{xx} - y^2 u_{yy} &= 0, \qquad (x, y > 0) \\ u(x,1) &= \varphi(x) \\ u_y(x,1) &= \psi(x), \end{aligned}$$

where φ, ψ are continuous on $\mathbb{R}$.

10.5 Mixed problem for the wave equation

Let $l > 0$ be a real number and we let $D =]0, +\infty[\times]0, l[$, further let $\rho, k, q : [0, l] \to \mathbb{R}$ be functions, where ρ, q are continuous, and k is continuously differentiable satisfying $\rho(x)k(x) > 0$ and $q(x) \geqslant 0$, whenever $0 \leqslant x \leqslant l$. Let $f : \overline{D} \to \mathbb{R}$, $\varphi, \psi : [0, l] \to \mathbb{R}$ and $\chi_1, \chi_2 : [0, +\infty[$ be continuous functions. Finally, for each function $u : D^{cl} \to \mathbb{R}$ function which is twice continuously differentiable on D, and continuously differentiable on D^{cl} we define

$$Lu(t,x) = \rho(x)\partial_t^2 u(t,x) - \partial_x\big(k(x)\partial_x u(t,x)\big) + q(x)u(t,x),$$

whenever (t, x) is in D. By a *mixed problem* for the equation

$$Lu = f(t,x) \tag{10.11}$$

we mean to find a function u with the above properties which is a solution of (10.11) on D and it satisfies

$$u(0,x) = \varphi(x), \qquad \partial_t u(0,x) = \psi(x) \tag{10.12}$$

and

$$u(t,0) = \chi_1(t), \qquad u(t,l) = \chi_2(t) \tag{10.13}$$

whenever $0 \leqslant x \leqslant l$ and $t \geqslant 0$. Here (10.11) is the so-called *wave equation*, the equations (10.12) are called *initial conditions*, and the equations (10.13) are called *boundary conditions*. The problem (10.11), (10.12), (10.13) is called a *mixed problem for the wave equation*. We note that the problem (10.11), (10.12), (10.13) is the problem of an inhomogeneous vibrating string of length l with varying endpoints under external forces, or shortly, *the problem of vibrating string*. In the case of homogeneous string equation (10.11) is replaced by

$$\partial_t^2 u - a^2\, \partial_x^2 u = f(t,x)\,,$$

where a is a constant. Here ρ is the density function of the string, k depends on the material properties of the string, and f is the external force. If χ_1 és χ_2 are continuously differentiable, then the obvious necessary conditions for the resolvability of the problem are the following:

$$\varphi(0) = \chi_1(0), \qquad \psi(0) = {\chi_1}'(0)\,,$$

$$\varphi(l) = \chi_2(0), \qquad \psi(l) = {\chi_2}'(0)\,,$$

the so-called *fitting conditions*.

Theorem 10.5.1. *The mixed problem* (10.11), (10.12), (10.13) *has at most one twice continuously differentiable solution.*

Proof. It is enough to show that in the case $f = 0$, $\varphi = \psi = 0$, $\chi_1 = \chi_2 = 0$ the only solution is $u = 0$. Hence we assume that u is a solution with these side conditions, and for $t \geqslant 0$ let

$$E(t) = \frac{1}{2}\int_0^l \left[\rho(x)\big(\partial_t u(t,x)\big)^2 + k(x)\big(\partial_x u(t,x)\big)^2 + q(x)\big(u(t,x)\big)^2\right] dx\,.$$

The function E is the so-called *energy integral*, the *total energy of the string*. Obviously $E(0) = 0$, and, on the other hand, for $t > 0$ we have $E'(t) = 0$, hence $E(t) = 0$ whenever $t \geqslant 0$. We conclude

$$\partial_t u(t,x) = 0, \qquad \partial_x u(t,x) = 0$$

for each (t,x) in D^{cl}, hence $u(t,x) = u(0,x) = 0$. □

Now we reduce the general mixed problem to four special ones. These are the following problems:

$$(A)\qquad \begin{aligned} &Lu = 0\,, \\ &u(0,x) = \varphi(x), \qquad \partial_t u(0,x) = \psi(x)\,, \\ &u(t,0) = 0, \qquad u(t,l) = 0\,, \end{aligned}$$

$$(B)\qquad \begin{aligned} &Lu = 0\,, \\ &u(0,x) = 0, \qquad \partial_t u(0,x) = 0\,, \\ &u(t,0) = \chi_1(t), \qquad u(t,l) = 0\,, \end{aligned}$$

$$(C)\qquad \begin{aligned} &Lu = 0\,, \\ &u(0,x) = 0, \qquad \partial_t u(0,x) = 0\,, \\ &u(t,0) = 0, \qquad u(t,l) = \chi_2(t)\,, \end{aligned}$$

$$(D)\qquad \begin{aligned} &Lu = f(t,x)\,, \\ &u(0,x) = 0, \qquad \partial_t u(0,x) = 0\,, \\ &u(t,0) = 0, \qquad u(t,l) = 0\,. \end{aligned}$$

Theorem 10.5.2. *Let the functions Φ_A, Φ_B, Φ_C, Φ_D be solutions of the mixed problems $(A), (B), (C), (D)$, respectively. Then $\Phi_A + \Phi_B + \Phi_C + \Phi_D$ is a solution of the mixed problem* (10.11), (10.12), (10.13).

Proof. The statement is obvious. □

The statement of this theorem is the *superposition principle* of waves. The following theorem provides further reduction.

Theorem 10.5.3. *Let in* (10.13) *the functions χ_1, χ_2 twice continuously differentiable, further we assume that*

$$g(t,x) = \chi_1(t) + \frac{x}{l}\big(\chi_2(t) - \chi_1(t)\big)$$

for each (t,x) in D^{cl}. Then Φ is a solution of the mixed problem (10.11), (10.12), (10.13) *if and only if $\Phi - g$ is a solution of the mixed problem*

$$\begin{aligned} &Lv = f(t,x) - Lg(t,x)\,, \\ &v(0,x) = \varphi(x) - g(0,x), \qquad \partial_t u(0,x) = \psi(x) - \partial_t g(0,x)\,, \\ &v(t,0) = 0, \qquad v(t,l) = 0\,. \end{aligned}$$

Proof. The statement is obvious. □

10.6 Problems

Solve the following mixed problems.

1.

$$\begin{aligned} u_{tt} &= u_{xx} - 4u, \qquad (0 < x < 1) \\ u(t,0) = 0 &\quad u(t,1) = 0 \\ u_t(0,x) = x^2 - x &\quad u_t(0,x) = 0 \end{aligned}$$

2.

$$\begin{aligned} u_{tt} + 2u_t &= u_{xx} - u, \qquad (0 < x < \pi) \\ u(t,0) = 0 &\quad u(t,\pi) = 0 \\ u_t(0,x) = \pi x - x^2 &\quad u_t(0,x) = 0 \end{aligned}$$

3.

$$\begin{aligned} u_{tt} &= u_{xx} + x, \qquad (0 < x < \pi) \\ u(t,0) = 0 &\quad u(t,\pi) = 0 \\ u_t(0,x) = \sin 2x &\quad u_t(0,x) = 0 \end{aligned}$$

4.

$$\begin{aligned} u_{tt} - u_{xx} + 2u_t &= 4x + 8e^t \cos t, \qquad \left(0 < x < \frac{\pi}{2}\right) \\ u(t, \frac{\pi}{2}) = \pi t &\quad u_x(t,0) = 2t \\ u(0,x) = \cos x &\quad u_t(0,x) = 2x \end{aligned}$$

10.7 Fourier method

We consider the mixed problem (A) in the previous section. Let

$$L_t u = \partial_t^2 u, \qquad L_x u = \frac{1}{\rho}\left(\partial_x(k\partial_x u) - qu\right),$$

then the first equation of the problem can also be written in the form $L_t u = L_x u$. First we are looking for the solutions of the mixed problem (A) of the form

$$u(t,x) = T(t) \cdot X(x).$$

Assume that u is a solution of (A), and for some point (x_0, y_0) in D we have $u(x_0, y_0) \neq 0$. Then, by substitution, we have

$$L_x X = \lambda X, \qquad X(0) = 0, \qquad X(l) = 0,$$

and

$$L_t T = \lambda T\,.$$

We have two eigenvalue problems. It is easy to see that all eigenvalues of the operator L_t are real numbers, and it is not difficult to compute its eigenfunctions. Concerning the eigenvalue problem for the operator L_x we can easily verify the following statements:

(1) Every eigensubspace is one-dimensional.
(2) Different eigensubspaces are orthogonal with respect to the weight function ρ.
(3) Every eigenvalue is negative.
(4) There are countably many eigenvalues and their only accumulation point can be $-\infty$.

In the space of all square integrable functions L^2_ρ with respect to the weight function ρ we consider the usual norm

$$\|X\|_\rho = \left(\int_0^l |X(x)|^2 \rho(x)\,dx\right)^{\frac{1}{2}},$$

and we suppose that all different eigenvalues of L_x are

$$0 > \lambda_1 > \lambda_2 > \cdots > \lambda_n > \cdots,$$

further the respective normed eigenfunctions are $X_1, X_2, \ldots, X_n, \ldots$. As λ_n is negative for each n, hence the general solution of the differential equation

$$L_t T - \lambda_n T = 0$$

is

$$T_n(t) = A_n \cos\sqrt{-\lambda_n}t + B_n \sin\sqrt{-\lambda_n}t\,.$$

Consequently, we let

$$U_n(t,x) = T_n(t)\cdot X_n(x)\,,$$

and

$$u(t,x) = \sum_{n=1}^{\infty} U_n(t,x)\,,$$

assuming that this series is uniformly convergent. Supposing that the summation and L commute we check under what assumptions will the function u satisfy the initial conditions. We get

$$\varphi(x) = u(0,x) = \sum_{n=1}^{\infty} A_n X_n(x)\,,$$

which implies

$$A_n = \int_0^l \varphi(x) X_n(x) \rho(x)\, dx$$

for $n = 1, 2, \ldots$. Similarly,

$$\psi(x) = \partial_t u(0, x) = \sum_{n=1}^{\infty} B_n \sqrt{-\lambda_n} X_n(x)\,,$$

hence

$$B_n = \frac{1}{\sqrt{-\lambda_n}} \int_0^l \psi(x) X_n(x) \rho(x)\, dx\,,$$

for $n = 1, 2, \ldots$. Finally, we have to check if the function

$$\Phi(t, x) = \sum_{n=1}^{\infty} \left(\int_0^l \varphi(\xi) X_n(\xi) \rho(\xi)\, d\xi \, \cos \sqrt{-\lambda_n} t \right.$$

$$\left. + \frac{1}{\sqrt{-\lambda_n}} \int_0^l \psi(\xi) X_n(\xi) \rho(\xi)\, d\xi \, \sin \sqrt{-\lambda_n} t \right) X_n(x)$$

is twice continuously differentiable and if it satisfies the mixed problem (A).

We note that similar method can be applied for the mixed problem (D) in the previous section.

10.8 Problems

Solve the following mixed problems.

1.

$$\begin{aligned} u_{tt} &= u_{xx} + u_{yy}, \qquad (0 < x < \pi, 0 < y < \pi) \\ u(t, 0, y) &= 0 \qquad u_x(t, \pi, y) = 0 \\ u(t, x, 0) &= 0 \qquad u_x(t, x, \pi) = 0 \\ u(0, x, y) &= 3 \sin x \sin 2y \qquad u_t(0, x, y) = 5 \sin 3x \sin 4y \end{aligned}$$

2.

$$\begin{aligned} u_t &= u_{xx} - 2u_x + x + 2t, \qquad (0 < x < 1) \\ u(t, 0) &= 0,\ u_x(t, 1) = t,\ u(0, x) = e^x \sin \pi x \end{aligned}$$

3.

$$u_t = u_{xx} + u - x + 2\sin 2x\cos x, \qquad \left(0 < x < \frac{\pi}{2}\right)$$
$$u(t,0) = 0,\ u_x\left(t, \frac{\pi}{2}\right) = 1,\ u(0,x) = x$$

4.

$$u_t = u_{xx} + 4u + x^2 - 2t - 4x^2t + 2\cos^2 x, \qquad (0 < x < \pi)$$
$$u_x(t,0) = 0,\ u_x(t,\pi) = 2\pi t,\ u(0,x) = 0$$

5.

$$u_t = u_{xx} - 2u_x + u + e^x\sin x - t\cos^2 x, \qquad (0 < x < \pi)$$
$$u(t,0) = 1 + t,\ u_x(t,\pi) = 1 + t,\ u(0,x) = 1 + e^x\sin 2x$$

Chapter 11

TABLE OF LAPLACE TRANSFORMS

$f(t)$	$\mathcal{L}[f(t)] = F(s)$
1	$\frac{1}{s}$
$e^{at}f(t)$	$F(s-a)$
$t^n f(t)$	$(-1)^n \frac{d^n F(s)}{ds^n}$
$f'(t)$	$sF(s) - f(0)$
$f^n(t)$	$s^n F(s) - s^{(n-1)} f(0) - \cdots - f^{(n-1)}(0)$
$\int_0^t f(x)g(t-x)dx$	$F(s)G(s)$
$t^n \ (n = 0, 1, 2, \dots)$	$\frac{n!}{s^{n+1}}$
$t^x \ (x \geqslant -1 \in \mathbb{R})$	$\frac{\Gamma(x+1)}{s^{x+1}}$

$\sin kt$	$\dfrac{k}{s^2+k^2}$
$\cos kt$	$\dfrac{s}{s^2+k^2}$
e^{at}	$\dfrac{1}{s-a}$
$\sinh kt$	$\dfrac{k}{s^2-k^2}$
$\cosh kt$	$\dfrac{s}{s^2-k^2}$
$\dfrac{e^{at}-e^{bt}}{a-b}$	$\dfrac{1}{(s-a)(s-b)}$
$\dfrac{ae^{at}-be^{bt}}{a-b}$	$\dfrac{s}{(s-a)(s-b)}$
te^{at}	$\dfrac{1}{(s-a)^2}$
$t^n e^{at}$	$\dfrac{n!}{(s-a)^{n+1}}$
$e^{at}\sin kt$	$\dfrac{k}{(s-a)^2+k^2}$
$e^{at}\cos kt$	$\dfrac{s-a}{(s-a)^2+k^2}$
$e^{at}\sinh kt$	$\dfrac{k}{(s-a)^2-k^2}$
$e^{at}\cosh kt$	$\dfrac{s-a}{(s-a)^2-k^2}$

$t \sin kt$	$\dfrac{2ks}{(s^2+k^2)^2}$
$t \cos kt$	$\dfrac{s^2-k^2}{(s^2+k^2)^2}$
$t \sinh kt$	$\dfrac{2ks}{(s^2-k^2)^2}$
$t \cosh kt$	$\dfrac{s^2-k^2}{(s^2-k^2)^2}$
$\dfrac{\sin at}{t}$	$\arctan \dfrac{a}{s}$
$\dfrac{1}{\sqrt{\pi t}} e^{-a^2/4t}$	$\dfrac{e^{-a\sqrt{s}}}{\sqrt{s}}$
$\dfrac{a}{2\sqrt{\pi t^3}} e^{-a^2/4t}$	$e^{-a\sqrt{s}}$

Chapter 12

ANSWERS TO SELECTED PROBLEMS

1.2 Basic concepts and terminology

1.2.6. Let $\varphi : I \to \mathbb{R}$ be a solution of the given Cauchy problem on the open interval I, then we have

$$\frac{d}{dt}\varphi(t)\cdot e^{-1} = \varphi'(t)e^{-1} - \varphi(t)e^{-t} = 0\,,$$

hence $t \mapsto \varphi(t)\cdot e^{-}$ is constant on I. It follows that $\varphi(t) = ce^{-1}$ on I with some constant c. By uniqueness, our statement follows.
1.2.9.i) $x = \frac{1}{2}(t+C)^2$
1.2.9.ii) $x = \frac{1}{2}\ln(2t+C),\ t > -\frac{C}{2}$
1.2.9.iii) $x = \tan(t+C),\ -\frac{\pi}{2} - C + k\pi < t < \frac{\pi}{2} - C + k\pi,\ k$ is an integer
1.2.9.iv) $\ln|x| + \frac{x^2}{2} = t + C$
1.2.9.v) $\int e^{-x^2}\,dx = t + C$

1.10 Parametric differential equations

1.10.1. $e^{2x} - x - 1$
1.10.2. $\frac{x^5}{5} - \frac{2x^3}{3} + x + 1$
1.10.3. $t(e^{-1} - e^{-t})$
1.10.4. t^8

1.12 Characteristic function

1.12.1.i) e^{x-2}
1.12.1.ii) $\frac{1-t-\ln(1-t)}{(1-t)^2}$
1.12.1.iii) $t^2\ln t + 2t^2 - 2t$
1.12.1.iv) $-e^{2t} - 2e^{-t} - 3e^{-2t}$

1.12.2.i) $\Phi(t, \tau, \xi) = \xi\left(\frac{t+1}{\tau+1}\right)^2$
1.12.2.ii) $\Phi(t, \tau, \xi) = \frac{\xi\tau}{(1-\xi)x+\xi\tau}$

2.2 Separable differential equations

2.2.1.i) $y = C(x+1)e^{-x}$
2.2.1.ii) $\ln|x| = C + \sqrt{y^2+1}$
2.2.1.iii) $y^2 - 2 = Ce^{1/x}$
2.2.1.iv) $(Ce^{-x^2} - 1)y = 2$
2.2.2. $e^y = \frac{x^3}{3} + 2$
2.2.3. $x^2 - 4x + y^3 - 5y = 9$
2.2.4.i) $y^2 - 4y = 2x^3 + 2x^2 + 2x + C$
2.2.4.ii) $\ln|\sin y| = \cos x + C, y \equiv k\pi (k = 0, \pm 1, \pm 2, \dots)$
2.2.4.iii) $y = \frac{C}{x-C}, y \equiv -1$
2.2.4.iv) $\frac{(\ln y)^2}{2} = -\frac{x^3}{3} + C$
2.2.4.v) $y^3 + 3\sin y + \ln|y| + \ln(1+x^2) + \tan^{-1} x = C, y \equiv 0$
2.2.4.vi) $y^2 = 1 + \left(\frac{x}{1+Cx}\right)^2$
2.2.5.i) $y = \tan\left(\frac{x^3}{3} + C\right)$
2.2.5.ii) $y = \frac{C}{\sqrt{1+x^2}}$
2.2.5.iii) $y = \frac{2-Ce^{(x-1)^2/2}}{1-Ce^{(x-1)^2/2}}, y \equiv 1$
2.2.5.iv) $y = 1 + (3x^2 + 9x + C)^{1/3}$
2.2.6.i) $y = 2 + \sqrt{2(x^3 + x^2 + x + 1}$
2.2.6.ii) $\frac{y}{y+1} = \frac{1}{2}e^{2-\frac{x^2}{2}}$

2.4 Differential equations of homogeneous degree

2.4.1.i) $x + y = Cx^2$
2.4.1.ii) $\ln(x^2 + y^2) = C - 2\tan^{-1}\frac{y}{x}$
2.4.1.iii) $x(y - x) = Cy, y \equiv 0$
2.4.1.iv) $x^2 = y^2 \ln Cx, y \equiv 0$
2.4.1.v) $y = Ce^{y/x}$
2.4.1.vi) $y^2 - x^2 = Cy, y \equiv 0$
2.4.2.i) $(y - 2x)^3 = C(y - x - 1)^2, y = x + 1$
2.4.2.ii) $2x + y - 1 = Ce^{2y-x}$
2.4.2.iii) $(y - x + 2)^2 + 2x = C$
2.4.2.iv) $(y - x + 5)^5(x + 2y - 2) = C$
2.4.2.v) $(y + 2)^2 = C(x + y - 1), y = 1 - x$

2.4.2.vi) $y + 2 = Ce^{-2\tan^{-1}\frac{y+2}{x-3}}$
2.4.2.vii) $\ln\frac{y+x}{x+3} = 1 + \frac{C}{x+y}$

2.6 First order linear differential equations

2.6.1.i) $y = Cx^2 + x^4$
2.6.1.ii) $y = (2x+1)(C + \ln|2x+1|) + 1$
2.6.1.iii) $y = \sin x + C\cos x$
2.6.1.iv) $y = e^x(\ln|x| + C)$
2.6.1.v) $xy = C - \ln|x|$
2.6.1.vi) $y = x(C + \sin x)$
2.6.1.vii) $y = Ce^{x^2} - x^2 - 1$
2.6.1.viii) $y = C\ln^2 x - \ln x$
2.6.1.ix) $xy = (x^3 + C)e^{-x}$
2.6.2.i) $y = \frac{e^{-(x-1)}}{x}$
2.6.2.ii) $y = \frac{e}{x\ln x}$
2.6.2.iii) $y = \frac{\pi}{x\sin x}$
2.6.2.iv) $y = 2(1+x^2)$
2.6.3.i) $y = \frac{1}{3} + Ce^{-3x}$
2.6.3.ii) $y = \frac{2}{x} + \frac{C}{x}e^x$
2.6.3.iii) $y = e^{-x^2}\left(\frac{x^2}{2} + C\right)$
2.6.3.iv) $y = -\frac{e^{-x}+C}{1+x^2}$
2.6.3.v) $y = (x + C)\cos x$
2.6.3.vi) $y = \frac{C-\cos x}{(1+x)^2}$
2.6.3.vii) $y = -\frac{1}{2}\frac{(x-2)^3}{x-1} + C\frac{(x-2)^5}{x-1}$
2.6.3.viii) $y = (x + C)e^{-\sin^2 x}$
2.6.3.ix) $y = \frac{e^x}{x^2} - \frac{e^x}{x^3} + \frac{C}{x^2}$
2.6.4.i) $y = \frac{1}{10}e^{-7x}\left(e^{10x} - 1\right)$
2.6.4.ii) $y = \frac{2x+1}{(x^2+1)^2}$
2.6.4.iii) $y = \frac{\ln(1+x^2)-\ln 2}{x^3}$
2.6.4.iv) $y = \frac{1}{2}(\sin x + \csc x)$
2.6.4.v) $y = \frac{2\ln|x|}{x} + \frac{x}{2} - \frac{1}{2x}$

2.8 Bernoulli equations

2.8.1.i) $u^2 = 6t^2 + Ct^{4/3}$
2.8.1.ii) $u = \frac{2e^t}{C-e^{2t}}$

2.8.1.iii) $u^3 = \frac{C}{t^3} + 1$
2.8.1.iv) $y = \frac{1}{1-Ce^x}$
2.8.1.v) $y = x^{2/7}(C - \ln|x|)^{1/7}$
2.8.1.vi) $y = e^{2/x}(C - \frac{1}{x})^2$
2.8.1.vii) $y^2 = \frac{2x+C}{(1+x^2)^2}$
2.8.1.viii) $y^2 = \frac{1}{1-x^2+Ce^{-x^2}}$
2.8.1.ix) $y^3 = \frac{x}{3(1-x)+Ce^{-x}}$
2.8.2.i) $y = \frac{2\sqrt{2}}{\sqrt{1-4x}}$
2.8.2.ii) $y = \left[1 - \frac{3}{2}e^{-(x^2-1)/4}\right]^{-2}$
2.8.2.iii) $y = \frac{1}{x(11-3x)^{1/3}}$
2.8.2.iv) $y = (2e^x - 1)^2$
2.8.2.v) $y = (2e^{12x} - 1 - 12x)^{1/3}$

2.10 Riccati equations

2.10.1.i) $y = 1 + \frac{1}{x+1+Ce^x}$
2.10.1.ii) $y = e^x - \frac{1}{1+Ce^{-x}}$
2.10.1.iii) $y = 1 - \frac{1}{x(1-Cx)}$
2.10.1.iv) $y = x - \frac{2x}{x^2+C}$
2.10.2.i) $y_p = e^{-x}$, $y = \frac{e^{-x}}{1+Ce^x}$
2.10.2.ii) $y_p = \frac{1}{(1+x)\ln(1+x)}$, $y = \frac{1}{(1+x)\left(C+\ln(1+x)\right)}$
2.10.2.iii) $y_p = -\sqrt[3]{3x^2}$, $y^3 = (Cx-3)x^2$
2.10.2.iv) $y_p = \sqrt{-2x}$, $y^2 = x(Cx-2)$
2.10.2.v) $y_p = \sqrt{\frac{1}{2e^x x^4}}$, $y^2 = \frac{1}{2e^x x^4 + Cx^4}$

2.12 Exact differential equations

2.12.1.i) $2x^3y^2 = C$
2.12.1.ii) $x^3 - 3y = C$
2.12.1.iii) $3y(\sin x + 1) + 2x^2e^x = C$
2.12.1.vi) $x^3 + 3xy^2 = C$
2.12.1.vii) $x^2 - 2xy^2 + 4y^3 = C$
2.12.1.viii) $\rho(1 + e^{2\theta}) = C$
2.12..1.ix) $x + y = C$
2.12.1.x) $x^2 + 3xy + 2y^2 + 4x + 5y = C$
2.12.1.xii) $2y^2\cos x + 3xy^3 - x^2 = C$

2.12.1.xiv) $x^3 + x^2y + 4xy^2 + 9y^2 = C$
2.12.2.i) $x^4y^2 - 2x^3y - x^2 - 3x + 1 = 0$
2.12.2.ii) $2y^2 - 4y\sin x - \cos 2x + \tan x = 1$
2.12.2.iii) $y^3 = \frac{e^x-1}{e^x+1}$
2.12.2.iv) $y\cos x = 2\sin x + \cos x$
2.12.2.v) $x^2y - x^2 - xy + x - 6y = 6$
2.12.3.i) $M(x,y) = 2xy + C(x)$
2.12.3.ii) $M(x,y) = 2(y\sin y + \cos y)(x\cos x + \sin x) + C(x)$
2.12.3.iii) $M(x,y) = ye^x - e^y\cos x + C(x)$
2.12.4.i) $N(x,y) = \frac{1}{2}x^4y + x^2 + 6xy + C(y)$
2.12.4.ii) $N(x,y) = \frac{x}{y} - 2\cos x$
2.12.4.iii) $N(x,y) = x\sin y + xy\cos y$
2.12.5.i) $\mu(x) = \frac{1}{x^2}$
2.12.5.ii) $\mu(x) = x^{-3/2}$
2.12.5.iii) $\mu(y) = y^{-3}$
2.12.5.iv) $\mu(x) = e^{\frac{5}{2}x}$
2.12.6.i) $\mu(x,y) = x^4y^3,\ \ x^5y^4\ln|x| = C$
2.12.6.ii) $\mu(x,y) = x^{-2}y^{-3},\ \ 3x^2y^2 + y = 1 + Cxy^2, y \equiv 0$
2.12.6.iii) $\mu(x,y) = x^{-2}y^{-1},\ \ -\frac{2}{x} + y^3 + 3\ln|y| = C, y \equiv 0$
2.12.6.iv) $\mu(x,y) = x^{-4}y^{-3},\ \ xy = C$
2.12.6.v) $\mu(x,y) = xe^y, x^2ye^y\sin x = C$

2.14 Incomplete differential equations

2.14.1.

$$\begin{aligned} x(p) &= \frac{1}{p^3} + \frac{1}{p^2} \\ y(p) &= \frac{3}{2p^3} + \frac{2}{p} + C \end{aligned}$$

2.14.2.

$$\begin{aligned} x(t) &= t - \frac{1}{t^2} \\ y(t) &= t - \frac{1}{2t^2} + \frac{2}{5t^5} + C \end{aligned}$$

2.14.3.

$$\begin{aligned} x(t) &= -t + \frac{1}{3}\ln|3t-1| + C \\ y(t) &= \frac{3}{t} - \frac{1}{t^2} \end{aligned}$$

2.14.4.

$$\begin{aligned} x(p) &= pe^p + e^p + C \\ y(p) &= p^2 e^p \end{aligned}$$

2.14.5.

$$\begin{aligned} x(t) &= \frac{1}{t} + C \\ y(t) &= \frac{1}{t} - t \end{aligned}$$

2.14.6.

$$\begin{aligned} x(p) &= -2\tan^{-1} p - \frac{2p}{1+p^2} + C \\ y(p) &= \frac{2}{1+p^2} \end{aligned}$$

2.16 Implicit differential equations

2.16.1.

$$\begin{aligned} x(p) &= p^3 + p \\ 4y(p) &= 3p^3 + 2p^2 + C \end{aligned}$$

2.16.2.

$$\begin{aligned} x(p) &= \frac{2p}{p^2 - 1} \\ y(p) &= \frac{2}{p^2 - 1} - \ln|p^2 - 1| + C \end{aligned}$$

2.16.3.

$$\begin{aligned} x(p) &= p\sqrt{1+p^2} \\ 3y(p) &= (2p^2 - 1)\sqrt{1+p^2} + C \end{aligned}$$

2.16.4.

$$\begin{aligned} x(p) &= \ln p + \frac{1}{p} \\ y(p) &= p - \ln p + C \end{aligned}$$

2.16.5. $y = 0$, or

$$\begin{aligned} x(p) &= 3p^2 + 2p + C \\ y(p) &= 2p^3 + p^2 \end{aligned}$$

2.16.6.

$$y = 0, \text{ or}$$
$$x(p) = 2\tan^{-1} p + C$$
$$y(p) = \ln(1 + p^2)$$

2.16.7.

$$y = \pm 1, \text{ or}$$
$$x(p) = \ln|p| \pm \frac{3}{2} \ln\left|\frac{\sqrt{p+1} - 1}{\sqrt{p+1} + 1}\right| \pm 3\sqrt{p+1} + C$$
$$y(p) = p \pm (p+1)^{3/2}$$

2.16.8.

$$y = -1, \text{ or}$$
$$x(p) = e^p + C$$
$$y(p) = (p-1)e^p$$

2.16.9.

$$y = 0, \text{ or}$$
$$x(p) = \pm\left(2\sqrt{p^2 - 1} + \sin^{-1}\frac{1}{|p|}\right) + C$$
$$y(p) = \pm p\sqrt{p^2 - 1}$$

2.16.10.

$$y = 0, \text{ or}$$
$$x(p) = \pm\frac{3}{2}\ln\left|\frac{1 - \sqrt{1-p}}{1 + \sqrt{1-p}}\right| + 3\sqrt{1-p} + C$$
$$y(p) = p \pm p\sqrt{1-p}$$

2.16.11.

$$y = 0, \text{ or}$$
$$x(p) = \pm 2\sqrt{1+p^2} - \ln(\sqrt{1+p^2} \pm 1) + C$$
$$y(p) = -p \pm \sqrt{1+p^2}$$

2.16.12.

$$4y = C^2 - 2(x - C)^2;\ 2y = x^2$$

2.16.13.

$$x^2 = 4y, \text{ or}$$
$$x(p) = -\frac{p}{2} + C$$
$$5y(p) = C^2 - \frac{5p^2}{4}$$

2.16.14.

$$\pm x(p)\sqrt{2\ln Cp} = 1$$
$$y(p) = \mp\left(\sqrt{2\ln Cp} - \frac{1}{\sqrt{2\ln Cp}}\right)$$

2.16.15.
$$y = 0, \text{ or}$$
$$px(p)y(p) = y(p)^2 + p^3$$
$$y(p)^2(2p + C) = p^4$$

2.16.16.
$$y^2 = 2Cx - C\ln C;\ 2x = 1 + 2\ln|y|$$

2.16.17.
$$Cx = \ln Cy;\ y = ex$$

2.16.18.
$$y = 0, \text{ or}$$
$$x(p)p^2 = C\sqrt{|p|} - 1$$
$$y(p) = x(p)p - x(p)^2p^3$$

2.18 Lagrange and Clairut equations

2.18.1.
$$y = Cx - C^2;\ 4y = x^2$$

2.18.2.
$$y = 0, \text{ or}$$
$$x\sqrt{p} = \ln p + C$$
$$y = \sqrt{p}(4 - \ln p - C)$$

2.18.3.
$$y = 0, \text{ or}$$
$$x = 3p^2 + Cp^{-2}$$
$$y = 2p^3 - 2Cp^{-1}$$

2.18.4.
$$y = Cx - C - 2$$

2.18.5.
$$C^3 = 3(Cx - y);\ 9y^2 = 4x^3$$

2.18.6.
$$y = 0, \text{ or } y = x - 2, \text{ or}$$
$$x = C(p-1)^{-2} + 2p + 1$$
$$y = Cp^2(p-1)^{-2} + p^2$$

2.18.7.
$$y = Cx - \ln C;\ y = \ln x + 1$$

2.18.8.
$$y = \pm 2\sqrt{Cx} + C;\ y = -x$$

2.18.9.
$$2C^2(y - Cx) = 1;\ 8y^3 = 27x^2$$

2.18.10.
$$xp^2 = p + C$$
$$y = 2 + 2Cp^{-1} - \ln p$$

3.2 Integrals of linear differential equations

3.2.1. $R(s,t) = \begin{bmatrix} 1 & s-t \\ 0 & 1 \end{bmatrix}$

3.2.2. $R(s,t) = \begin{bmatrix} \left(\frac{s}{t}\right)^2 & 0 \\ 0 & \left(\frac{s}{t}\right)^3 \end{bmatrix}$

3.4 Linear differential equations with constant coefficients

3.4.1.i)

$$\begin{aligned} x &= (C_1 + C_2 t)e^{2t} + 3 \\ y &= (C_1 + C_2 + C_2 t)e^{2t} + 1 \end{aligned}$$

3.4.1.ii)

$$\begin{aligned} x &= C_1 e^t(\cos 3t - 2\sin 3t) + C_2 e^t(2\cos 3t + \sin 3t) \\ y &= C_1 e^t(\cos 3t + \sin 3t) + C_2 e^t(\sin 3t - \cos 3t) \end{aligned}$$

3.4.1.iii)

$$\begin{aligned} x &= C_1 + [C_2 + C_3(t+1)]e^{-t} \\ y &= [C_2 + C_3(t-1)]e^{-t} \\ z &= C_1 + (C_2 + C_3 t)e^{-t} \end{aligned}$$

3.4.2.i) $y = C_1 \begin{bmatrix} 1 \\ 1 \\ 0 \end{bmatrix} e^t + C_2 \begin{bmatrix} 1 \\ 0 \\ 0 \end{bmatrix} e^{-t} + C_3 \begin{bmatrix} 1 \\ -2 \\ 1 \end{bmatrix} e^{-2t}$

3.4.2.ii) $y = C_1 \begin{bmatrix} -1 \\ 0 \\ 1 \end{bmatrix} e^{-2t} + C_2 \begin{bmatrix} 0 \\ -1 \\ 1 \end{bmatrix} e^{-2t} + C_3 \begin{bmatrix} 1 \\ 1 \\ 1 \end{bmatrix} e^{4t}$

3.4.2.iii) $y = C_1 \begin{bmatrix} -1 \\ 0 \\ 1 \end{bmatrix} e^{-3t} + C_2 \begin{bmatrix} 0 \\ -1 \\ 1 \end{bmatrix} e^{-3t} + C_3 \begin{bmatrix} 1 \\ 1 \\ 1 \end{bmatrix} e^{3t}$

3.4.2.iv) $y = C_1 \begin{bmatrix} 1 \\ 0 \\ 1 \end{bmatrix} e^{2t} + C_2 \begin{bmatrix} 1 \\ -1 \\ 1 \end{bmatrix} e^{3t} + C_3 \begin{bmatrix} 1 \\ -3 \\ 7 \end{bmatrix} e^{-t}$

3.4.3.i) $y = -\begin{bmatrix} 2 \\ 6 \end{bmatrix} e^{5t} + \begin{bmatrix} 4 \\ 2 \end{bmatrix} e^{-5t}$

3.4.3.ii) $y = \begin{bmatrix} 2 \\ -4 \end{bmatrix} e^{t/2} + \begin{bmatrix} -2 \\ 1 \end{bmatrix} e^{t}$

3.4.3.iii) $y = \begin{bmatrix} 7 \\ 7 \end{bmatrix} e^{9t} - \begin{bmatrix} 2 \\ 4 \end{bmatrix} e^{-3t}$

3.4.3.iv) $y = \begin{bmatrix} 3 \\ 9 \end{bmatrix} e^{5t} - \begin{bmatrix} 4 \\ 2 \end{bmatrix} e^{-5t}$

3.4.3.v) $y = \begin{bmatrix} 5 \\ 5 \\ 0 \end{bmatrix} e^{t/2} + \begin{bmatrix} 0 \\ 0 \\ 1 \end{bmatrix} e^{t/2} + \begin{bmatrix} -1 \\ 2 \\ 0 \end{bmatrix} e^{-t/2}$

3.4.3.vi) $y = \begin{bmatrix} 3 \\ 3 \\ 3 \end{bmatrix} e^{t} + \begin{bmatrix} -2 \\ -2 \\ 2 \end{bmatrix} e^{-t}$

3.4.3.vii) $y = \begin{bmatrix} 2 \\ -2 \\ 2 \end{bmatrix} e^{t} - \begin{bmatrix} 3 \\ 0 \\ 3 \end{bmatrix} e^{-2t} + \begin{bmatrix} 1 \\ 1 \\ 0 \end{bmatrix} e^{3t}$

3.4.4.i) $y = C_1 \begin{bmatrix} 2 \\ 1 \end{bmatrix} e^{5t} + C_2 \Big(\begin{bmatrix} -1 \\ 0 \end{bmatrix} e^{5t} + \begin{bmatrix} 2 \\ 1 \end{bmatrix} te^{5t} \Big)$

3.4.4.ii) $y = C_1 \begin{bmatrix} 1 \\ 1 \end{bmatrix} e^{-t} + C_2 \Big(\begin{bmatrix} 1 \\ 0 \end{bmatrix} e^{-t} + \begin{bmatrix} 1 \\ 1 \end{bmatrix} te^{-t} \Big)$

3.4.4.iii) $y = C_1 \begin{bmatrix} -2 \\ 1 \end{bmatrix} e^{-9t} + C_2 \Big(\begin{bmatrix} -1 \\ 0 \end{bmatrix} e^{-9t} + \begin{bmatrix} -2 \\ 1 \end{bmatrix} te^{-9t} \Big)$

3.4.4.iv) $y = C_1 \begin{bmatrix} -1 \\ 1 \end{bmatrix} e^{2t} + C_2 \Big(\begin{bmatrix} -1 \\ 0 \end{bmatrix} e^{2t} + \begin{bmatrix} -1 \\ 1 \end{bmatrix} te^{2t} \Big)$

3.4.4.v) $y = C_1 \begin{bmatrix} -1 \\ 1 \\ 1 \end{bmatrix} e^{t} + C_2 \begin{bmatrix} 1 \\ -1 \\ 1 \end{bmatrix} e^{-t} + C_3 \Big(\begin{bmatrix} 0 \\ 3 \\ 0 \end{bmatrix} e^{-t} + \begin{bmatrix} 1 \\ -1 \\ 1 \end{bmatrix} te^{-t} \Big)$

3.4.4.vi) $y = C_1 \begin{bmatrix} 0 \\ 1 \\ 1 \end{bmatrix} e^{2t} + C_2 \begin{bmatrix} 1 \\ 0 \\ 1 \end{bmatrix} e^{-2t} + C_3 \Big(\frac{1}{2} \begin{bmatrix} 1 \\ 1 \\ 0 \end{bmatrix} e^{-2t} + \begin{bmatrix} 1 \\ 0 \\ 1 \end{bmatrix} te^{-t} \Big)$

3.4.4.vii) $y = C_1 \begin{bmatrix} -2 \\ -3 \\ 1 \end{bmatrix} e^{2t} + C_2 \begin{bmatrix} 0 \\ -1 \\ 1 \end{bmatrix} e^{4t} + C_3 \Big(\frac{1}{2} \begin{bmatrix} 1 \\ 0 \\ 0 \end{bmatrix} e^{4t} + \begin{bmatrix} 0 \\ -1 \\ 1 \end{bmatrix} te^{4t} \Big)$

3.4.4.viii) $y = C_1 \begin{bmatrix} -1 \\ -1 \\ 1 \end{bmatrix} e^{-2t} + C_2 \begin{bmatrix} 1 \\ 1 \\ 1 \end{bmatrix} e^{4t} + C_3 \Big(\frac{1}{2} \begin{bmatrix} 1 \\ 0 \\ 0 \end{bmatrix} e^{4t} + \begin{bmatrix} 1 \\ 1 \\ 1 \end{bmatrix} te^{4t} \Big)$

3.4.5.i) $y = \begin{bmatrix} -2 \\ 2 \\ 2 \end{bmatrix} e^{4t} + \begin{bmatrix} 0 \\ -1 \\ 1 \end{bmatrix} e^{2t} + \begin{bmatrix} 3 \\ -3 \\ 3 \end{bmatrix} te^{2t}$

3.4.5.ii) $y = -\begin{bmatrix} 1 \\ 1 \\ 0 \end{bmatrix} e^{-4t} + \begin{bmatrix} -3 \\ 2 \\ -3 \end{bmatrix} e^{8t} + \begin{bmatrix} 8 \\ 0 \\ -8 \end{bmatrix} te^{8t}$

3.4.5.iii) $y = \begin{bmatrix} 3 \\ 6 \\ 3 \end{bmatrix} e^{4t} - \begin{bmatrix} 3 \\ 4 \\ 1 \end{bmatrix} + \begin{bmatrix} 8 \\ 4 \\ 4 \end{bmatrix} t$

3.6 Computation of the exponential matrix

3.6.1.i)

$$e^{tA} = \begin{pmatrix} 1-2t & -4t \\ t & 1+2t \end{pmatrix}$$

3.6.1.ii)

$$e^{tA} = \begin{pmatrix} e^t(\cos t - 2\sin t) & 5e^t \sin t \\ -e^t \sin t & e^t(\cos t + 2\sin t) \end{pmatrix}$$

$$A = \begin{pmatrix} \operatorname{ch} t & 0 & \operatorname{sh} t \\ 0 & 1 & 0 \\ \operatorname{sh} t & 0 & \operatorname{ch} t \end{pmatrix}$$

$$A = \begin{pmatrix} e^{2t} & 0 & -te^{2t} \\ 0 & e^{2t} & 0 \\ 0 & 0 & e^{2t} \end{pmatrix}$$

3.6.2.

$$e^{\frac{\pi}{3}A} b = \begin{pmatrix} 1-3\sqrt{3} \\ 2-\sqrt{3} \end{pmatrix}$$

3.6.3.i)

$$A = \begin{pmatrix} 1+2t & -t \\ 4t & 1-2t \end{pmatrix}$$

3.6.3.ii)

$$A = \begin{pmatrix} e^t(1+2t) & 4te^t \\ -te^t & 1-2t \end{pmatrix}$$

3.6.3.iii)

$$A = \begin{pmatrix} \cos t + \sin t & -2\sin t \\ \sin t & \cos t - \sin t \end{pmatrix}$$

3.6.4.i)

$$y_1 = c_1e^{6x} - c_2e^{-2x}$$
$$y_2 = c_1e^{6x} + c_2e^{-2x}$$

3.6.4.ii)

$$y_1 = c_1e^{2x} + c_2e^{3x} + c_3e^{-x}$$
$$y_2 = -c_2e^{3x} - 3c_3e^{-x}$$
$$y_3 = c_1e^{2x} + c_2e^{3x} + 7c_3e^{-x}$$

3.6.4.iii)

$$y_1 = c_1e^{x} - c_2e^{-5x} - c_3e^{-5x}$$
$$y_2 = c_1e^{x} + c_3e^{-5x}$$
$$y_3 = c_1e^{x} + c_2e^{-5x}$$

3.6.4.i)

$$y_1 = 2c_1e^{2x} - c_2\sin x + c_3\cos x$$
$$y_2 = 2c_1e^{2x} - c_2(\cos x + \sin x) + c_3(\cos x - \sin x)$$
$$y_3 = c_1e^{2x} + c_2\cos x + c_3\sin x$$

4.4 First integrals

4.4.1. $\frac{1}{2}x^2 - y = C$
4.4.2. $y^2 + z^2 = C_1, \;\; x - yz = C_2$
4.4.3. $\frac{z-x}{y-x} = C_1, \;\; (x-y)^2(x+y+z) = C_2$
4.4.4. $\frac{zy}{x} = C_1, \;\; x^2 + y^2 + z^2 = C_2$
4.4.5. $xy = C_1, \;\; (x+y)(x+y+z) = C_2$
4.4.6. $x = \ln(C_1t + C_2), \;\; y = \ln(C_1t + C_2) + C_3 - C_2, \;\; z = (C_1 + 1)t + C_2$

5.4 Intermediate integrals

5.4.1.

$$x = \sin t$$
$$y = -\frac{1}{8}t\cos 2t = \frac{3}{16}t + C_1\sin t + \frac{7}{48}\sin 2t + C_2\cos 2t - \frac{1}{192}\sin 4t + C_3$$

5.4.2. $x + C_2 = \frac{2}{3}(\sqrt{y} + C_1)^{3/2} - 2C_1(\sqrt{y} + C_1)^{1/2}$
5.4.3. $y = \frac{1}{6a^3C^5}[C^4(x + C_1)^2 - a^6]^{3/2} - \frac{a^3}{2C^3}(x + C_1)\ln[C^2(x + C_1)\sqrt{C^4(x + C_1)^2 - a^6}] + \frac{a^3}{2C^5}\sqrt{C^4(x + C_1)^2 - a^6} + C_2x + C_3$

5.4.4. $y = \operatorname{sh}(x + C_1) + C_2 x + C_3$
5.4.5.

$$x = C_1(t - \sin t) + C_2$$
$$y = C_1(1 - \cos t) + 2a$$

5.4.6. $y = \frac{C_1 x^3}{6} - \frac{C_1^3 x^3}{3} + C_2 x + C_3$
5.4.7. $\ln y = C_1 e^x + C_2 e^{-x}$

5.6 Higher order linear differential equations

5.6.1. $W(x) = \frac{C}{1-x^2}$
5.6.2. $y = C_1 \frac{\sin x}{x} + C_2 \frac{\cos x}{x}$
5.6.3. $y = C_2 + (C_1 - C_2 x) \cot x$
5.6.4. $y = C_1 x + C_2 x^2 + C_3 x^3$
5.6.5. $y = C_1 x + C_2 \sin x + C_3 \cos x$
5.6.6. $y = C_1 + C_2 x^2 + C_3(\sin^{-1} x + x\sqrt{1 - x^2})$
5.6.7. $y = C_1 x + C_2 x^2 + x^3$
5.6.8. $y = C_1 e^x + C_2 x - (x^2 + 1)$
5.6.9. $y = C_1 \cos x + C_2 \sin x + C_3 x^2 + x^3$

5.8 Linear differential equations with constant coefficients

5.8.1.i) $y = C_1 e^x + C_2 e^{-2x}$
5.8.1.ii) $y = C_1 e^{-x} + C_2 e^{-3x}$
5.8.1.iii) $y = C_1 + C_2 e^{2x}$
5.8.1.iv) $y = C_1 e^{2x} + C_2 e^{x/2}$
5.8.1.v) $y = e^{2x}(C_1 \sin x + C_2 \cos x)$
5.8.1.vi) $y = e^{-x}(C_1 \sin 3x + C_2 \cos 3x)$
5.8.1.vii) $y = e^x(C_1 + C_2 x + C_3 x^2)$
5.8.1.viii) $y = e^x(C_1 + C_2 x) + C_3 e^{-x}$
5.8.1.ix) $y = C_1 e^x + C_2 e^{-x} + C_3 e^{2x} + C_4 e^{-2x}$
5.8.1.x) $y = C_1 + (C_2 + C_3 x) \cos x + (C_4 + C_5 x) \sin x$
5.8.2.i) $y = e^x(x \ln|x| + C_1 x + C_2)$
5.8.2.ii) $y = (e^{-x} + e^{-2x}) \ln(e^x + 1) + C_1 e^{-x} + C_2 e^{-2x}$
5.8.2.iii) $y = (C_1 + \ln|\sin x|) \sin x + (C_2 - x) \cos x$
5.8.2.iv) $y = \sin 2x \ln|\cos x| - x \cos 2x + C_1 \sin 2x + C_2 \cos 2x$
5.8.2.v) $y = e^{-x}\left(\frac{4}{5}(x+1)^{5/2} + C_1 + C_2 x\right)$
5.8.2.vi) $y = C_1 \cos x + C_2 \sin x - \frac{\cos 2x}{\cos x}$

5.8.3.i) $y = (7 - 3x)e^{x-2}$
5.8.3.ii) $y = 2\cos x - 5\sin x + 2e^x$
5.8.3.iii) $y = e^{2x-1} - 2e^x + e - 1$
5.8.3.iv) $y = e^{-x}(x - \sin x)$
5.8.3.v) $y = 2 + e^{-x}$
5.8.3.vi) $y = (x-1)(e^{2x} - e^{-x})$
5.8.3.vii) $y = x - x\sin x - 2\cos x$

5.10 Decreasing the order of linear homogeneous equations

5.10.1. $y = C_1 x + C_2 x^2 + C_3 x^3$
5.10.2. $y = C_1 x + C_2 \sin x + C_3 \cos x$
5.10.3. $y = C_1 + C_2 x^2 + C_3(\arcsin x + x\sqrt{1 - x^2})$

5.12 Euler differential equations

5.12.1. $y = C_1 x^2 + C_2 x^3 - \frac{1}{2}x$
5.12.2. $y = x(C_1 \cos\ln x + C_2 \sin\ln x) - x\ln x$
5.12.3. $y = \frac{C_1}{x} + \frac{1}{3}\left[\left(x^2 - \frac{1}{x}\right)\ln x - \frac{1}{3x} - \frac{x^2}{3}\right]$
5.12.4. $y = x(C_1 + C_2 \ln x) + C_3 x^2 + \frac{1}{4}x^3 - \frac{3}{2}x(\ln x)^2$
5.12.5. $y = C_1 \cos\ln(1+x) + [C_2 + 2\ln(1+x)]\sin\ln(1+x)$

5.14 Exponential polynomials

5.14.1. $y = e^{2x}(C_1 + C_2 x) + \frac{x^2}{4} + \frac{x}{2} + \frac{3}{8}$
5.14.2. $y = C_1 e^{2x} + C_2 e^{4x} + \frac{1}{3}e^x - \frac{x}{2}e^{2x}$
5.14.3. $y = C_1 e^{-x} + C_2 \cos x + C_3 \ln x + e^x\left(\frac{x}{4} - \frac{3}{8}\right)$
5.14.4. $y = e^x\left(C_1 + C_2 x + C_3 x^2 + C_4 x^3 + \frac{1}{24}x^4 + \frac{1}{120}x^5\right)$
5.14.5. $y = C_1 \cos 2x + C_2 \sin 2x - \frac{x^2}{8}\cos 2x + \frac{x}{16}\sin 2x$

5.16 Boundary value problems

5.16.1. $p = \pi^2$
5.16.2. $a = 2$
5.16.3.i)

$$G(x,s) = \begin{cases} (s-1)x & \text{if } 0 \leqslant x \leqslant s \\ s(x-1) & \text{if } s \leqslant x \leqslant 1 \end{cases}$$

5.16.3.ii)

$$G(x,s)=\begin{cases}\sin s\cos x & \text{if } 0\leqslant x\leqslant s\\ \cos s\sin x & \text{if } s\leqslant x\leqslant \pi\end{cases}$$

5.16.3.iii)

$$G(x,s)=\begin{cases}e^{s}(e^{-x}-1) & \text{if } 0\leqslant x\leqslant s\\ 1-e^{s} & \text{if } s\leqslant x\leqslant 1\end{cases}$$

5.16.3.iv)

$$G(x,s)=\begin{cases}-e^{-s}\operatorname{ch} x & \text{if } 0\leqslant x\leqslant s\\ -e^{-x}\operatorname{ch} s & \text{if } s\leqslant x\leqslant 2\end{cases}$$

5.16.3.v)

$$G(x,s)=\begin{cases}\frac{1}{x}-1 & \text{if } 0\leqslant x\leqslant s\\ \frac{1}{s}-1 & \text{if } s\leqslant x\leqslant 3\end{cases}$$

5.16.3.vi)

$$G(x,s)=\begin{cases}\frac{s^2-4}{2s^2} & \text{if } 0\leqslant x\leqslant s\\ \frac{x^2-4}{2s^2} & \text{if } s\leqslant x\leqslant 2\end{cases}$$

5.16.3.vii)

$$G(x,s)=\begin{cases}\frac{1-x^3}{3s^3x} & \text{if } 0\leqslant x\leqslant s\\ \frac{1-s^3}{3s^3x} & \text{if } s\leqslant x\leqslant 2\end{cases}$$

5.16.4.i) $\lambda_k=-k^2\pi^2/l^2,\ \ y_k=\sin k\pi x/l,\ \ k=1,2,\dots$
5.16.4.ii) $\lambda_k=-k^2\pi^2/l^2,\ \ y_k=\cos k\pi x/l,\ \ k=0,1,\dots$
5.16.4.iii) $\lambda_k=-\left(k-\frac{1}{2}\right)^2\frac{\pi^2}{l^2},\ \ y_k=\sin\left(k-\frac{1}{2}\right)\frac{\pi x}{l},\ \ k=1,2,\dots$
5.16.4.iv) $\lambda_k=-\left(\frac{k\pi}{\ln a}\right)^2-\frac{1}{4},\ \ y_k=\sqrt{x}\sin\frac{k\pi\ln x}{\ln a},\ \ k=1,2,\dots$
5.16.5.i) $y=\frac{\operatorname{sh} x}{\operatorname{sh} 1}-2x$
5.16.5.ii) $y=x+e^{-x}-e^{-1}$
5.16.5.iii) $y=e^{x}-2$
5.16.5.iv) $y=1-\sin x-\cos x$
5.16.5.v) There is no solution.
5.16.5.vi) $y=2x-\pi+\pi\cos x+C\sin x,$ C is an arbitrary constant

5.20 Power series solutions

5.20.1. $y=x$
5.20.2. $y=c_1x+c_2\sqrt[3]{x}$
5.20.3. $y=c_1(1-3x^2)+c_2(x-\frac{1}{3}x^3)$
5.20.4. $y=x^2e^{-x}$

5.22 The Laplace transform

5.22.1.i) $F(s) = \frac{6}{s+5} + \frac{1}{s-3} + \frac{30}{s^4} - \frac{9}{s}$
5.22.1.ii) $F(s) = \frac{4s}{s^2+16} - \frac{36}{s^2+16} + \frac{2s}{s^2+100}$
5.22.1.iii) $F(s) = \frac{6}{s^2-4} + \frac{6}{s^2+4}$
5.22.1.iv) $F(s) = \frac{1}{s-3} + \frac{s}{s^2+36} - \frac{s-3}{(s-3)^2+36}$
5.22.2.i) $F(s) = \frac{1}{s-3} + \frac{s}{s^2+36} - \frac{s-3}{(s-3)^2+36}$
5.22.2.ii) $F(s) = \frac{12s^2-16}{(s^2+4)^3}$
5.22.2.iii) $F(s) = \frac{3\sqrt{\pi}}{4s^{\frac{5}{2}}}$
5.22.2.iv) $F(s) = 10^{\frac{3}{2}} \frac{3\sqrt{\pi}}{4s^{\frac{5}{2}}}$
5.22.3.i) $f(x) = 6 - e^{8x} + 4e^{3x}$
5.22.3.ii) $f(x) = 19e^{-2x} - \frac{1}{3}e^{\frac{5x}{3}} + \frac{7}{24}x^4$
5.22.3.iii) $f(x) = 6\cos 5x + \frac{3}{5}\sin 5x$
5.22.3.iv) $f(x) = \frac{4}{3}\sin 2x + \frac{3}{7}\sinh 7x$
5.22.4.i) $f(x) = 6\cos\sqrt{7}x - \frac{5}{\sqrt{7}}\sin\sqrt{7}x$
5.22.4.ii) $f(x) = -3e^{-4x}\cos\sqrt{5}x + \frac{13}{\sqrt{5}}e^{-4x}\sin\sqrt{5}x$
5.22.4.iii) $f(x) = \frac{1}{2}\left(3e^{\frac{3x}{2}}\cosh\frac{\sqrt{5}}{2}x + \frac{5}{\sqrt{13}}e^{\frac{3x}{2}}\sinh\frac{\sqrt{5}}{2}x\right)$
5.22.4.iv) $f(x) = -\frac{5}{7}e^{-2x} + \frac{12}{7}e^{5x}$
5.22.5.i) $f(x) = -3e^{-3x} + +2e^{4x} + e^{\frac{x}{5}}$
5.22.5.ii) $f(x) = \frac{1}{47}\left(-28e^{6x} + 28\cos\sqrt{11}x - \frac{67}{\sqrt{11}}\sin\sqrt{11}x\right)$
5.22.5.iii) $f(x) = \frac{1}{5}\left(11 - 20x - \frac{25}{2}x^2 - 11e^{-2x}\cos x - 2e^{-2x}\sin x\right)$
5.22.6.i) $y = \frac{50}{81} + \frac{5}{9}x + \frac{31}{81}e^{9x} - 2e^x$
5.22.6.ii) $y = \frac{1}{125}\left(-96e^{\frac{x}{2}} + 96e^{-2x} - 10xe^{-2x} - \frac{25}{2}x^2e^{-2x}\right)$
5.22.6.iii) $y = \frac{1}{10}\left(\cos 3x + \frac{1}{3}\sin 3x - 11e^{3x}\cos\sqrt{6}x - \frac{8}{\sqrt{6}}\sin\sqrt{6}x\right)$

5.24 The Fourier transform of exponential polynomials

5.24.1.i) $y(x) = (C_1x^2 + C_2x + C_3)e^x + \frac{1}{4}\left(-x\cos x + (x+3)\sin x\right)$
5.24.1.ii) $y(x) = C_1\cos 2x + C_2\sin 2x - \frac{x^2}{8}\cos 2x + \frac{x}{16}\sin 2x$
5.24.1.iii) $y(x) = 2xe^x(x-1) + C_1x + C_2e^x + C_3e^{-x}$
5.24.1.iv) $y(x) = x^2\ln|x| + C_1x + C_2x^2 + \frac{C_3}{x} + \frac{C_4}{x^2}$
5.24.1.v) $y(x) = 2x^3 + C_1x + C_2x^2 + C_3xe^x$
5.24.1.vi) $y(x) = x\ln|x| + C_1x + C_2x^2 + C_3x^3$
5.24.2.i) $y(x) = C_1x + C_2x\ln x$
5.24.2.ii) $y(x) = -\frac{e^{-x}(x+1)}{x} + C_1e^x + C_2\frac{e^x}{x} + C_3\frac{e^{-x}}{x}$
5.24.2.iii) $y(x) = \frac{x^2\ln|x|}{3} + C_1x + C_2x^2 + C_3\frac{1}{x}$

5.24.2.iv) $y(x) = -x^2 - 2 + C_1x + C_2e^x + C_3e^{-x}$

6.2 Homogeneous linear partial differential equations

6.2.1.i) $u(x,y) = f(x+y)$
6.2.1.ii) $u(x_1, x_2, \dots, x_n) = f\left(\frac{x_2}{x_1}, \frac{x_3}{x_1}, \dots, \frac{x_n}{x_1}\right)$
6.2.1.iii) $u(x,y) = f(x^2+y^2)$
6.2.2. $u(x,y) = \frac{x}{y}$
6.2.3. $f(x,y,z) = \varphi(\sqrt{x}-\sqrt{y}, \sqrt{x}-\sqrt{z})\,;\, f(x,y,z) = (1-\sqrt{x}+\sqrt{y})^2 - (1-\sqrt{x}+\sqrt{z})^2$
6.2.4. $u(x,y,z) = x^2+y^2+z^2 - \frac{y^2}{x^2}$
6.2.5. $u(x,y) = \sqrt{x^2+y^2-1}$
6.2.6.i) $u(x,y) = f\left(\frac{x}{y}\right)$
6.2.6.ii) $u(x,y,z,) = f(xz-y, z)$
6.2.6.iii) $u(x,y,z) = f\left(\frac{x}{y}, x^2+y^2+z^2\right)$
6.2.6.iv) $u(x,y) = f\left(\ln x + \frac{1}{y}\right)$
6.2.6.v) $u(x,y,z) = f\left(\frac{x}{y}, x^2+y^2-2z\right)$
6.2.7.i) $u(x,y) = f(x^2+y^2), f(1) = 1$
6.2.7.ii) $u(x,y) = x-y$
6.2.7.iii) $u(x,y) = \sqrt{xy}$

6.4 Quasilinear partial differential equations

6.4.1.i) $F\big((x-u)S^{1/3}, (y-u)S^{1/3}, (z-u)S^{1/3}\big) = 0$, where $S = x+y+z+u$
6.4.1.ii) $u = \frac{1}{2}x^2yz - \frac{1}{6}x^3(bz+cy) + \frac{1}{12}bcx^4 + f(y-bx, z-cx)$
6.4.1.iii) $u = \frac{1}{x^3y^3} f\left(\frac{x}{y^2} + \frac{y}{x^2}\right)$
6.4.1.iv) $z = \sqrt{xy}$.
6.4.2.i) $u = x^a f\left(\frac{y}{x}\right)$
6.4.2.ii) $u^2 = e^{x^2-y^2} f(x^2+y^2)$
6.4.2.iii) $u = \frac{y^2+3xf(xy)}{3x}$
6.4.2.iv) $u = \frac{1}{12}\Big(abx^4 - 2bx^3y - 2ax^3z + 6x^2yz\Big) + f(y-ax, z-bx)$
6.4.2.v) $F\big(x^2+y^2+u^2, cx+ay+bu\big) = 0$
6.4.3.i) $u = 2xe^{-y} - x - 2ye^{-y} - y$
6.4.3.ii) $u = \frac{1}{2}\Big(x^4 + x^2y^2 - 2\sqrt{x^2+y^2}\Big)$
6.4.3.iii) $u = \sqrt{x^2+y^2}$
6.4.3.iii) $u = (x+y)z$
6.4.3.iv) $u = \sqrt{\frac{y^4}{x^2} + 2(x^2+y^2)\ln x}$

6.4.3.v) $u = y$

7.2 First order partial differential equations

7.2.1.

$$\begin{aligned} x' &= x - q \\ y' &= y - p \\ u' &= -pq \\ p' &= -p \\ q' &= -q \end{aligned}$$

7.2.1.

$$\begin{aligned} x' &= a \\ y' &= b \\ z' &= c \\ u' &= xyz \\ p' &= -yz \\ q' &= -xz \\ r' &= -xy \end{aligned}$$

7.2.3.

$$\begin{aligned} x' &= q \\ y' &= p \\ u' &= 2u \\ p' &= p \\ q' &= q \end{aligned}$$

7.2.4.

$$\begin{aligned} x' &= 2p \\ y' &= 2q \\ u' &= 2 \\ p' &= 0 \\ q' &= 0 \end{aligned}$$

7.2.5.

$$\begin{aligned} x' &= x \\ y' &= y \\ u' &= u - xy \\ p' &= -y \\ q' &= -x \end{aligned}$$

7.4 Cauchy problem for first order equations

7.4.1. $u(x,y) = y(\sqrt{x^2-1}+x)$
7.4.2. $u(x,y) = \frac{1}{16}x^2 + \frac{1}{2}xy + y^2$
7.4.3.

$$\begin{aligned} x &= t + s \\ y &= -2st + s^2 \\ u &= -s^2t + s^3 \end{aligned}$$

7.4.4. $u(x,y) = y$
7.4.5. $u(x,y) = \frac{x}{2} + y$
7.4.6. $u(x,y) = \sqrt{2}(x+y) + 1$

7.6 Special Cauchy problem for first order partial differential equation

7.6.1. $y^2 - x^2 - \ln\sqrt{y^2-x^2} = z - \ln|y|$
7.6.2. $2x^2(y+1) = y^2 + 4z - 1$
7.6.3. $(x+2y)^2 = 2x(u+xy)$
7.6.4. $\sqrt{\frac{u}{y^3}}\sin x = \sin\sqrt{\frac{u}{y}}$
7.6.5. $2xy + 1 = x + 3y + u^{-1}$
7.6.6. $x - 2y = x^2 + y^2 + u$
7.6.7. $2x^2 - y^2 - u^2 = a^2$

7.8 Complete integral

7.8.1.i) $\Phi(x,y,a,b) = a\frac{x^2}{2} + \frac{1}{a}\frac{y^2}{2} + b$
7.8.1.ii) $\Phi(x,y,a,b) = x\cos a + y\sin a + b$
7.8.1.iii) $\Phi(x,y,a,b) = \frac{1}{4a}(x+ay+b)^2$
7.8.1.iv) $\Phi(x,y,a,b) = x - y + a\ln xy + b$

7.8.1.v) $\Phi(x, y, a, b) = -\frac{1}{a}\cos ax + ay + b$
7.8.1.vi) $\Phi(x, y, a, b) = \sqrt{a}x + \sqrt[3]{a}y + b$
7.8.1.vii) $\Phi(x, y, a, b) = \frac{a}{2}x + \frac{1}{a}y + b$
7.8.2.i) $u(x, y) = \sqrt{x^2 + y^2} - 1$
7.8.2.ii) $u(x, y) = xy + y\sqrt{x^2 + 1}$

8.6 Exponential polynomial solutions of partial differential equations

8.6.1.i) $u(t, x) = 1 + e^t + \frac{1}{2}t^2$
8.6.1.ii) $u(t, x) = t^3 + e^{-t}\sin x$
8.6.1.iii) $u(t, x) = (1 + t)e^{-t}\cos x$
8.6.1.iv) $u(t, x) = \cosh t \sin x$
8.6.2.i) $u(t, x, y) = e^t - 1 + e^{-2t}\cos x \sin y$
8.6.2.ii) $u(t, x, y) = 1 + \frac{1}{5}\sin x \sin y(2\sin t - \cos t + e^{-2t})$
8.6.3.i) $u(t, x, y, z) = \frac{1}{4}\cos x(e^{-2t} - 1 + 2t) + \cos x \cos z e^{-4t}$
8.6.3.ii) $u(t, x, y, z) = e^t - 1 + \sin(x - y - z)e^{-9t}$
8.6.4.i) $u(t, x) = x^4 + t^2\left(\frac{1}{2}x^3 - 12\right) + itx(12x + t^2)$
8.6.4.ii) $u(t, x, y) = x\sin t + x^2 + y^2\cos t + 2i(t + \sin t)$
8.6.4.iii) $u(t, x, y, z) = i(x^3 + y^3 + z^3) - t(6y + 6z - y^2 - iz^3) + t^2(i - 3z)$

9.2 Second order partial differential equations with linear principal part

9.2.1.i) $u(x, y) = \int_{x_0}^{x}\int_{y_0}^{y} f(\xi, \eta)\,d\eta\,d\xi + C_1(x) + C_2(y)$
9.2.1.ii) $u(x, y) = e^y\left[C_1(x) + \int_{x_0}^{x}\int_{y_0}^{y} f(\xi, \eta)e^{-\eta}\,d\eta\,d\xi\right] + C_2(y)$
9.2.1.iii) $u(x, y) = C_1(x + y) + C_2(x - y)$
9.2.1.iv) $u(x, y) = C_1(ax + y) + C_2(ax - y)$
9.2.1.v) $u(x, y) = \int_{x_0}^{x} C_1(\xi + y)\,d\xi + C_2(y)$
9.2.2.i) $u(x, y) = y + \frac{1}{2}\left[x^2y + xy^2 - x^3 - y^3\right]$
9.2.2.ii) $u(x, y) = 1 + x$
9.2.3.i) $A(x) = C_1e^{\sqrt{\lambda}x} + C_2e^{-\sqrt{\lambda}x}$, $\lambda > 0$, or $A(x) = C_1v + C_2$, or $A(x) = C_1\sin(\sqrt{-\lambda}x) + C_2\cos(\sqrt{-\lambda}x)$, $\lambda < 0$, and $B(y) = C_3e^{\lambda y}$
9.2.3.ii) $u(x, y) = C_1x + \sqrt{1 - C_1^2}y + C_2$
9.2.3.iii) $u(x, y) = C_1x + C_2y + \sqrt{1 - C_1^2 - C_2^2}z + C_3$

9.4 Linear transformation, normal form

9.4.1.i) $u_{\xi\xi} + u_{\eta\eta} + u_{\zeta\zeta} = 0$

$$\xi = x$$
$$\eta = y - x$$
$$\zeta = x - \frac{1}{2}y + \frac{1}{2}z$$

9.4.1.ii) $u_{\xi\xi} - u_{\eta\eta} + u_{\zeta\zeta} + u_{\eta} = 0$

$$\xi = \frac{1}{2}x$$
$$\eta = \frac{1}{2}x + y$$
$$\zeta = -\frac{1}{2}x - y + z$$

9.4.1.iii) $u_{\xi\xi} - u_{\eta\eta} + 2u_{\xi} = 0$

$$\xi = x + y$$
$$\eta = y - x$$
$$\zeta = y + z$$

9.4.1.iv) $u_{\xi\xi} - u_{\eta\eta} - u_{\zeta\zeta} = 0$

$$\xi = x$$
$$\eta = y - x$$
$$\zeta = \frac{3}{2}x - \frac{1}{2}y + \frac{1}{2}z$$

9.4.1.v) $u_{\xi\xi} + u_{\eta\eta} + u_{\zeta\zeta} + u_{\tau\tau} = 0$

$$\xi = x$$
$$\eta = y - x$$
$$\zeta = x - y + z$$
$$\tau = 2x - 2y + z + t$$

9.4.1.vi) $u_{\xi\xi} - u_{\eta\eta} + u_{\zeta\zeta} + u_{\tau\tau} = 0$

$$\xi = x + y$$
$$\eta = y - x$$
$$\zeta = z$$
$$\tau = y + z + t$$

9.4.1.vii) $u_{\xi\xi} - u_{\eta\eta} + u_{\zeta\zeta} + u_{\tau\tau} = 0$

$$\begin{aligned} \xi &= x + y \\ \eta &= x - y \\ \zeta &= -2y + z + t \\ \tau &= z - t \end{aligned}$$

9.4.1.viii) $u_{\xi\xi} - u_{\eta\eta} + u_{\zeta\zeta} = 0$

$$\begin{aligned} \xi &= x \\ \eta &= y - x \\ \zeta &= 2x - y + z \\ \tau &= x + z + t \end{aligned}$$

9.4.1.ix) $u_{\xi\xi} + u_{\eta\eta} = 0$

$$\begin{aligned} \xi &= x \\ \eta &= y \\ \zeta &= -x - y + z \\ \tau &= x - y + t \end{aligned}$$

9.4.2.i) $\sum_{k=1}^{n} u_{\xi_k\xi_k} = 0, \quad \xi_k = \sum_{l=1}^{k} x_l, \quad k = 1, 2, \ldots, n$
9.4.2.ii) $\sum_{k=1}^{n} (-1)^{k+1} u_{\xi_k\xi_k} = 0, \quad \xi_k = \sum_{l=1}^{k} x_l, \quad k = 1, 2, \ldots, n$
9.4.2.iii) $\sum_{k=1}^{n} u_{\xi_k\xi_k} = 0, \quad \xi_1 = x_1, \xi_k = x_k - x_{k-1} \quad k = 2, 3, \ldots, n$
9.4.2.iv) $\sum_{k=1}^{n} u_{\xi_k\xi_k} = 0\,,$

$$\xi_k = \sqrt{\frac{2k}{k+1}}\Big(x_k - \frac{1}{k}\sum_{l<k} x_l\Big), \quad k = 1, 2, \ldots, n$$

9.4.2.v) $u_{\xi_1\xi_1} - \sum_{k=2}^{n} u_{\xi_k\xi_k} = 0, \xi_1 = x_1 + x_2, \xi_2 = x_1 - x_2\,,$

$$\xi_k = \sqrt{\frac{2(k-2)}{k-1}}\Big(x_k - \frac{1}{k-2}\sum_{l<k} x_l\Big), \quad k = 3, 4, \ldots, n$$

9.6 Reduced normal form

9.6.1. $v_{\xi\eta} + (c - ab)v = 0$
9.6.2. $v_{\xi\eta} + 3v = 0$
9.6.3. $u_{\xi\xi} + u_{\eta\eta} = 0$

9.9 Hyperbolic equations

9.9.1. $u_{\xi\eta} - \frac{1}{16}(u_\xi - u_\eta) = 0$

$$\xi = x - y$$
$$\eta = 3x + y$$

9.9.2. $u_{\xi\eta} + \frac{1}{4(\eta^2-\xi^2)}(\eta u_\xi + \xi u_\eta) = 0$

$$\xi = y^2 - x^2$$
$$\eta = y^2 + x^2$$

9.9.3. $u_{\xi\eta} - \frac{1}{2(\xi-\eta)}(u_\xi - u_\eta) + \frac{1}{4(\xi+\eta)}(u_\xi + u_\eta) = 0$

$$\xi = y^2 + e^x$$
$$\eta = y^2 - e^x$$

9.11 Parabolic equations

9.11.1. $u_{\eta\eta} + u_\xi = 0$

$$\xi = x - 2y$$
$$\eta = x$$

9.11.2. $u_{\eta\eta} - u_\xi = 0$

$$\xi = \ln x + y$$
$$\eta = y$$

9.13 Elliptic equations

9.13.1.i) $u_{\xi\xi} + u_{\eta\eta} + u_\xi = 0$

$$\xi = x$$
$$\eta = 3x + y$$

9.13.1.ii) $u_{\xi\xi} + u_{\eta\eta} - u_\xi - u_\eta = 0$

$$\xi = \ln|x|$$
$$\eta = \ln|y| \qquad (x, y \neq 0)$$

9.13.1.iii) $u_{\xi\xi} + u_{\eta\eta} + \frac{1}{2\xi}u_\xi + \frac{1}{2\eta}u_\eta = 0$

$$\xi = y^2$$
$$\eta = x^2$$

9.13.1.iv) $u_{\xi\xi} + u_{\eta\eta} - \tanh \xi u_\xi = 0$

$$\xi = \ln(x + \sqrt{1+x^2})$$
$$\eta = \ln(y + \sqrt{1+y^2})$$

9.13.1.v) $u_{\xi\xi} + u_{\eta\eta} + \cos\xi \cdot u_\eta = 0$

$$\xi = x$$
$$\eta = y - \cos x$$

9.13.2.i) $u(x,y) = \sqrt{xy}C_1\left(\frac{y}{x}\right) + C_2(xy)$
9.13.2.ii) $u_{\xi\xi} + u_{\eta\eta} = 0$
9.13.2.iii) $u(x,y) = xC_1\left(\frac{y}{x}\right) + C_2\left(\frac{y}{x}\right)$
9.13.3.i) $u_{\eta\eta} - \frac{\xi}{\xi e^\eta + 1}u_\xi = 0$

$$\xi = e^{-x} - e^{-y}$$
$$\eta = y$$

9.13.3.ii) $u_{\xi\xi} + u_{\eta\eta} - u_\eta = 0$

$$\xi = x$$
$$\eta = \ln|1+y|, \qquad (y \neq -1)$$

9.13.3.iii) $u_{\xi\xi} + u_{\eta\eta} + \frac{1}{2\xi}u_\xi + \frac{1}{2\eta}u_\eta = 0$

$$\xi = y^2$$
$$\eta = x^2, \qquad (x, y \neq 0)$$

9.13.3.iv) $(\xi^2 - \eta^2)u_{\xi\eta} + (\xi - 3\eta + 2)u_\xi + (3\xi - \eta + 2)u_\eta = 0$

$$\xi = e^x + \frac{y^2}{2}$$
$$\eta = e^x - \frac{y^2}{2}, \qquad (y \neq 0)$$

9.13.3.v) $u_{\xi\eta} + \eta u_\xi + 4u_\eta = 0$

$$\xi = y - 3x$$
$$\eta = y - 2x$$

9.13.3.vi) For $y < 0$ we have $u_{\xi\eta} + \frac{1}{2(\xi-\eta)}(u_\xi - u_\eta) = 0$ with

$$\xi = x + 2\sqrt{-y}$$
$$\eta = x - 2\sqrt{-y}\,,$$

and for $y > 0$ we have $u_{\xi\xi} + u_{\eta\eta} - \frac{1}{\eta}u_\eta = 0$ with

$$\xi = x$$
$$\eta = 2\sqrt{y}\,.$$

9.13.3.vii) For $x < 0, y > 0$ we have $-12u_{\xi\eta} + \left(\frac{6}{\xi+\eta} + \frac{2}{\xi-\eta}\right)u_\xi + \left(\frac{6}{\xi+\eta} + \frac{2}{\xi-\eta}\right)u_\eta) = 0$ with

$$\begin{aligned}\xi &= 3\sqrt{y} + \sqrt{-x^3}\\ \eta &= 3\sqrt{y} - \sqrt{-x^3}\,,\end{aligned}$$

for $x > 0, y < 0$ we have $-12u_{\xi\eta} + \left(\frac{6}{\xi+\eta} + \frac{2}{\xi-\eta}\right)u_\xi + \left(\frac{6}{\xi+\eta} + \frac{2}{\xi-\eta}\right)u_\eta) = 0$ with

$$\begin{aligned}\xi &= 3\sqrt{-y} + \sqrt{x^3}\\ \eta &= 3\sqrt{-y} - \sqrt{x^3}\,,\end{aligned}$$

for $x > 0, y > 0$ we have $u_{\xi\xi} + u_{\eta\eta} + \frac{1}{\xi}u_\xi - \frac{1}{\eta}u_\eta = 0$ with

$$\begin{aligned}\xi &= \sqrt{x^3}\\ \eta &= 3\sqrt{y}\,,\end{aligned}$$

and for $x < 0, y < 0$ we have $u_{\xi\xi} + u_{\eta\eta} + \frac{1}{\xi}u_\xi - \frac{1}{\eta}u_\eta = 0$ with

$$\begin{aligned}\xi &= \sqrt{-x^3}\\ \eta &= 3\sqrt{-y}\,.\end{aligned}$$

9.13.4.i) $u(x,y) = C_1(x+y) + C_2(x-y)$
9.13.4.ii) $u(x,y) = C_1(x+y)e^{\frac{y-3x}{2}} + C_2(3x-y)$
9.13.4.iii) $u(x,y) = C_1(x+y)e^{\frac{y-x}{4}} + C_2(y-x)$
9.13.4.iv) $u(x,y) = C_1(\ln|y-1| - x) + C_2(y)$ in the open regions bounded by $y = x$ and $y = 1$
9.13.5.i) For $x > 0$ we have $u_{\xi\eta} + \frac{1}{6(\xi+\eta)}(u_\xi + u_\eta) = 0$ with

$$\begin{aligned}\xi &= \frac{2}{3}\sqrt{x^3} + y\\ \eta &= \frac{2}{3}\sqrt{x^3} + y\,,\end{aligned}$$

and for $x < 0$ we have $u_{\xi\xi} + u_{\eta\eta} - \frac{1}{3\xi}u_\xi = 0$ with

$$\begin{aligned}\xi &= \frac{2}{3}\sqrt{-x^3}\\ \eta &= y\end{aligned}$$

9.13.5.ii) For $y > 0$ we have $u_{\xi\eta} + \frac{1}{2(\xi-\eta)}(u_\xi - u_\eta) = 0$ with

$$\begin{aligned}\xi &= x + 2\sqrt{y}\\ \eta &= x - 2\sqrt{y}\,,\end{aligned}$$

and for $y < 0$ we have $u_{\xi\xi} + u_{\eta\eta} - \frac{1}{\eta}u_\eta = 0$ with

$$\begin{aligned}\xi &= x\\ \eta &= 2\sqrt{-y}\end{aligned}$$

9.13.5.iii) For $x > 0, y > 0$ or $x < 0, y < 0$ we have $u_{\xi\xi} - u_{\eta\eta} - \frac{1}{\xi}(u_\xi - u_\eta) = 0$ with

$$\xi = \sqrt{|x|}$$
$$\eta = \sqrt{|y|},$$

and for $x > 0, y < 0$ or $x < 0, y > 0$ we have $u_{\xi\xi} + u_{\eta\eta} - \frac{1}{\xi}(u_\xi + u_\eta) = 0$ with

$$\xi = \sqrt{|x|}$$
$$\eta = \sqrt{|y|}$$

9.13.5.iv) For $x > 0, y > 0$ or $x < 0, y < 0$ we have $u_{\xi\xi} - u_{\eta\eta} + \frac{1}{3\xi}u_\xi - \frac{1}{3\eta}u_\eta = 0$ with

$$\xi = \sqrt{|x|^3}$$
$$\eta = \sqrt{|y|^3},$$

and for $x > 0, y < 0$ or $x < 0, y > 0$ we have $u_{\xi\xi} + u_{\eta\eta} + \frac{1}{3\xi}u_\xi + \frac{1}{3\eta}u_\eta = 0$ with

$$\xi = \sqrt{|x|^3}$$
$$\eta = \sqrt{|y|^3}$$

9.13.6.i) $u(x, y) = f(x) + g(y)$
9.13.6.ii) $u(x, y) = f(y + ax) + g(y - ax)$
9.13.6.iii) $u(x, y) = f(x - y) + g(3x + y)$
9.13.6.iv) $u(x, y) = f(y) + g(x)e^{-ay}$
9.13.6.v) $u(x, y) = x - y + f(x - 3y) + g(2x + y)e^{\frac{3y-x}{7}}$
9.13.6.vi) $u(x, y) = [f(x) + g(y)]e^{-bx-ay}$
9.13.6.vii) $u(x, y) = e^{x+y} + [f(x) + g(y)]e^{3x+2y}$
9.13.6.viii) $u(x, y) = f(y - ax) + g(y - ax)e^{-x}$
9.13.7.i) For $x \neq -y$ we have $u(x, y) = f(x + y) + (x - y)g(x^2 - y^2)$
9.13.7.ii) For $x, y \neq 0$ we have $u(x, y) = f(xy) + \sqrt{|xy|}g(\frac{x}{y})$
9.13.7.iii) For $x, y \neq 0$ we have $u(x, y) = f(xy) + \sqrt[4]{|xy|^3}g(\frac{x}{y})$
9.13.7.iv) For $x^2 + y^2 \neq 0$ we have $u(x, y) = f(\frac{x}{y}) + xg(\frac{x}{y})$
9.13.7.v) $u(x, y) = xf(y) + f'(y) + \int_0^x (x - \xi)g(\xi)e^{\xi y}\, d\xi$ (Hint: with the notation $v = u_x$ show that $u = xv - v_y$, and $v_{xy} - xv_x = 0$.)
9.13.7.vi) $u(x, y) = 2yg(x) + \frac{1}{x}g'(x) + \int_0^y (y - \xi)f(\xi)e^{-x^2\xi}\, d\xi$ (Hint: with the notation $v = u_y$ show that $u = \frac{1}{2x}v_x + yv$, and $v_{xy} + 2xyv_y = 0$.)
9.13.7.vii) $u(x, y) = e^{-y}\left[yf(x) + f'(x) + \int_0^y (y - \eta)g(\eta)e^{-x\eta}\, d\eta\right]$ (Hint: with the notation $v = u + u_y$ show that $u = v_x + yv$, and $v_{xy} + v_x + yv_y + yv = 0$.)
9.13.7.viii) $u(x, y) = e^{-xy}\left[yf(x) + f'(x) + \int_0^y (y - \eta)g(\eta)e^{-x\eta}\, d\eta\right]$ (Hint: with the notation $v = xu + u_y$ show that $u = v_x + 2yv$, and $(v_y + xv)_x + 2y(v_y + xv) = 0$.)

10.2 Goursat problem for hyperbolic equations

10.2.1. $u(x,y) = y^2 + \frac{1}{2}x^2(1+e^{-y})$
10.2.2. $u(x,y) = \int_0^x e^{-\frac{1}{2}\xi^2 y^2}\, d\xi$
10.2.3. $u(x,y) = \left(1 + \frac{1}{2}x - \frac{1}{2}y\right)e^{\frac{1}{2}(x-y)}$
10.2.4. $u(x,y) = 1 + (x+2y)e^{\frac{1}{3}(y-x)}$
10.2.5. $u(x,y) = x^2 + (y - 1 + e^x)^2$
10.2.6. $u(x,y) = \frac{1}{16}(x^2 - y^2)^2 + xy$

10.4 Cauchy problem for hyperbolic equations

10.4.1. $u(x,y) = \varphi(x+y) - \frac{1}{2}e^{-\frac{x+y}{2}} \int_{x-y}^{x+y} [\varphi'(\xi) - \psi'(\xi)]\, e^{\frac{\xi}{2}}\, d\xi$
10.4.2. $u(x,y) = \frac{1}{2}\varphi(xy) + \frac{y}{2}\psi\left(\frac{x}{y}\right) + \frac{1}{4}\sqrt{xy}\int_{xy}^{\frac{x}{y}} \xi^{-\frac{3}{2}}[\varphi(\xi) - 2\psi(\xi)]\, d\xi$

10.6 Mixed problem for the wave equation

10.6.1. $u(x,y) = -\frac{8}{\pi^3}\sum_{k=0}^{\infty} \frac{\sin(2k+1)\pi x}{(2k+1)^3} \cos(\sqrt{(2k+1)^2\pi^2 + 4}\, t)$
10.6.2. $u(x,y) = -\frac{8e^{-t}}{\pi}\sum_{k=0}^{\infty} \frac{1}{(2k+1)^3}\left[\cos(2k+1)t + \frac{1}{2k+1}\sin(2k+1)t\right]\sin(2k+1)x$
10.6.3. $u(x,y) = \sin 2x \cos 2t + \sum_{k=1}^{\infty}(-1)^k \frac{2}{k^3}(1 - \cos kt) sinkx$
10.6.4. $u(x,y) = 2xt + (2e^t - e^{-t} - 3te^{-t})\cos x$

10.8 Fourier method

10.8.1. $u(t,x,y) = 3\cos\sqrt{5}t \sin x \sin 2y + \sin 5t \sin 3x \sin 4y$
10.8.2. $u(t,x) = xt + (\sin \pi x)\, e^{x - t - \pi^2 t}$
10.8.3. $u(t,x) = x + t\sin x + \frac{1}{8}(1 - e^{-8t})\sin 3x$
10.8.4. $u(t,x) = tx^2 + \frac{1}{4}(e^{4t} - 1) + t\cos 2x$

Bibliography

Arnold, I. V. (2006). *Ordinary differential equations*, Universitext (Springer-Verlag, Berlin).

Boyce, W. E. and DiPrima, R. C. (2001). *Elementary Differential Equations and Boundary Value Problems* (John Wiley and Sons, New York, Toronto, Singapore).

Coddington, E. A. and Levinson, N. (1955). *Theory of ordinary differential equations* (McGraw-Hill Book Company, Inc., New York-Toronto-London).

Filippov, A. F. (2000). *Problems in differential equations (in Russian)* (NIC, "Regular and chaotic dynamics", Moscow).

Filippov, A. F. (2007). *Introduction to the theory of differential equations (in Russian)* (KomKniga, Moscow).

Hartman, P. (1964). *Ordinary Differential Equations and Boundary Value Problems* (John Wiley and Sons, New York, London, Sidney).

Kamke, E. (1944). *Differentialgleichungen. Lösungsmethoden und Lösungen. Band I. Gewöhnliche Differentialgleichungen* (Akademische Verlagsgesellschaft, Leipzig).

Kamke, E. (1946). *Differentialgleichungen. Lösungsmethoden und Lösungen. Band II. Partielle Differentialgleichungen erster Ordnung für eine gesuchte Funktion* (Akademische Verlagsgesellschaft, Leipzig; J. W. Edwards, Ann Arbor, Mich.).

Logan, J. D. (2011). *A first course in differential equations*, Undergraduate Texts in Mathematics (Springer, New York).

Petrovski, I. G. (1966). *Ordinary differential equations* (Prentice-Hall, Inc., Englewood Cliffs, N.J.).

Pontryagin, L. S. (1962). *Ordinary differential equations* (Addison-Wesley Publishing Co., Inc., Reading, Mass.-Palo Alto, Calif.-London).

Schroers, B. J. (2011). *Ordinary differential equations*, African Institute of Mathematics Library Series (Cambridge University Press, Cambridge), ISBN 978-1-107-69749-2.

Stepanow, W. W. (1963). *Lehrbuch der Differentialgleichungen* (VEB Deutscher Verlag der Wissenschaften, Berlin).

Székelyhidi, L. (1985). Exponential polynomials and differential equations, *Publ.*

Math. Debrecen **32**, 1–2, pp. 105–109.

Székelyhidi, L. (1986). The Fourier transform of exponential polynomials, *Publ. Math. Debrecen* **33**, 1–2, pp. 13–20.

Székelyhidi, L. (2006). *Discrete spectral synthesis and its applications* (Springer Monographs in Mathematics).

Vladimirov, V. S. (1974). *Problems on Equations of mathematical physics (in Russian)* ("Nauka", Moscow).

Vladimirov, V. S. (1988). *Equations of mathematical physics (in Russian)* ("Nauka", Moscow).

Index